CONTENTS

STATEMENTS OF COMMITTEE MEMBERS

WITNESSES

QUESTIONS AND ANSWERS

SUBMISSIONS FOR THE RECORD

S. HRG. 109–222

TERRORISM: EMERGENCY PREPAREDNESS

HEARING

BEFORE THE

SUBCOMMITTEE ON TERRORISM, TECHNOLOGY AND HOMELAND SECURITY

OF THE

COMMITTEE ON THE JUDICIARY UNITED STATES SENATE

ONE HUNDRED NINTH CONGRESS

FIRST SESSION

OCTOBER 26, 2005

Serial No. J–109–46

Printed for the use of the Committee on the Judiciary

U.S. GOVERNMENT PRINTING OFFICE

25–378 PDF WASHINGTON : 2006

For sale by the Superintendent of Documents, U.S. Government Printing Office
Internet: bookstore.gpo.gov Phone: toll free (866) 512–1800; DC area (202) 512–1800
Fax: (202) 512–2250 Mail: Stop SSOP, Washington, DC 20402–0001

COMMITTEE ON THE JUDICIARY

ARLEN SPECTER, Pennsylvania, *Chairman*

ORRIN G. HATCH, Utah
CHARLES E. GRASSLEY, Iowa
JON KYL, Arizona
MIKE DEWINE, Ohio
JEFF SESSIONS, Alabama
LINDSEY O. GRAHAM, South Carolina
JOHN CORNYN, Texas
SAM BROWNBACK, Kansas
TOM COBURN, Oklahoma

PATRICK J. LEAHY, Vermont
EDWARD M. KENNEDY, Massachusetts
JOSEPH R. BIDEN, JR., Delaware
HERBERT KOHL, Wisconsin
DIANNE FEINSTEIN, California
RUSSELL D. FEINGOLD, Wisconsin
CHARLES E. SCHUMER, New York
RICHARD J. DURBIN, Illinois

DAVID BROG, *Staff Director*
MICHAEL O'NEILL, *Chief Counsel*
BRUCE A. COHEN, *Democratic Chief Counsel and Staff Director*

SUBCOMMITTEE ON TERRORISM, TECHNOLOGY AND HOMELAND SECURITY

JON KYL, Arizona, *Chairman*

ORRIN G. HATCH, Utah
CHARLES E. GRASSLEY, Iowa
JOHN CORNYN, Texas
MIKE DEWINE, Ohio
JEFF SESSIONS, Alabama
LINDSEY O. GRAHAM, South Carolina

DIANNE FEINSTEIN, California
EDWARD M. KENNEDY, Massachusetts
JOSEPH R. BIDEN, JR., Delaware
HERBERT KOHL, Wisconsin
RUSSELL D. FEINGOLD, Wisconsin
RICHARD J. DURBIN, Illinois

STEPHEN HIGGINS, *Majority Chief Counsel*
STEVEN CASH, *Democratic Chief Counsel*

TERRORISM: EMERGENCY PREPAREDNESS

WEDNESDAY, OCTOBER 26, 2005

U.S. SENATE,
SUBCOMMITTEE ON TERRORISM, TECHNOLOGY AND HOMELAND SECURITY, COMMITTEE ON THE JUDICIARY,
Washington, DC.

The Subcommittee met, pursuant to notice, at 10:32 a.m., in room SD–226, Dirksen Senate Office Building, Hon. Jon Kyl, Chairman of the Subcommittee, presiding.

Present: Senators Kyl, Cornyn, Feinstein, and Durbin.

OPENING STATEMENT OF HON. JON KYL, A U.S. SENATOR FROM THE STATE OF ARIZONA

Chairman KYL. Good morning and welcome to the Subcommittee on Terrorism, Technology, and Homeland Security of the Senate Judiciary Committee. We are pleased to have all of you here this morning for what I think is going to be an enlightening and very important hearing.

Let me begin by noting that Senator Feinstein and I will make opening statements. If Senator Cornyn arrives at an appropriate time, I will call upon him for an opening statement, too, and then we will go right to our witnesses. We have one panel today, but I suspect that that one panel will engage in a pretty complete and lively discussion and there is no time constraint here is the main point I wanted to make.

Hurricane Katrina exposed the weakness of our Nation's emergency preparedness. As reported in an October 20 Washington Post article, Homeland Security Secretary Michael Chertoff acknowledged that Hurricane Katrina overwhelmed FEMA, exposing major flaws in the nation's preparedness for terrorism and natural disasters. Secretary Chertoff vowed to reengineer U.S. preparedness.

We have learned a lot in the weeks since Hurricane Katrina. Today, this Subcommittee will focus on the question of whether we are prepared for a possible terrorist attack involving problems similar to those caused by the natural disaster in the Gulf Coast. A moderately sophisticated terrorist attack could easily replicate the type and amount of damage caused by this natural disaster, I believe, though I will ask you all whether that is, in fact, correct, and the response would be even more difficult to coordinate because we wouldn't have much time in terms of warning, if any, as to when or where such an attack might occur.

The objective of this hearing is to gain a better understanding of the types of terrorist attacks that could still take place, specifically those that could have an impact similar to Hurricane Katrina's, the key success factors in planning for and responding to an attack, the emergency preparedness of the Federal Government and how it should work with State and local authorities to respond effectively, and any existing shortfalls that need attention by State, local, and Federal authorities to improve readiness.

This Subcommittee on Terrorism, Technology, and Homeland Security has held numerous hearings about terrorist attack, such as attacks against seaports, attacks with biological weapons, and attacks against critical information infrastructure. Earlier this year, the Subcommittee held a hearing on the potentially devastating impact of an electromagnetic pulse explosion.

Today, the Subcommittee will examine what should be done to achieve an immediate, effective, and successful response to terrorist attacks. The Subcommittee will hear from five expert witnesses, one former Senator and member of the 9/11 Commission, a private sector expert, two State officials from California, and a scholar from the Brookings Institution. Senator Feinstein will introduce the two California witnesses in her opening statement and I will introduce the other members of the panel.

I will begin with former Senator Slade Gorton. He has served in public office for four decades, 18 of those years here in the U.S. Senate. Late in 2002, then-Majority Leader Trent Lott appointed him to the National Commission on Terrorist Attacks upon the United States, where he served with distinction and played a prominent role in formulating the final report. He will testify to the Commission's findings and warnings about preparedness for terrorist attacks.

Wayne Thomas is Vice President of Homeland Security for Innovative Emergency Management, IEM, a Louisiana-based private corporation focused on improving emergency preparedness at Federal, State, and local levels. IEM has particular expertise planning for responses to natural disasters and attacks involving weapons of mass destruction. Founded in 1985, IEM has worked with Federal organizations such as the Office of Domestic Preparedness, the Federal Emergency Management Agency, and the Department of Defense, as well as State and local emergency management agencies in more than 25 States. Before joining IEM, Mr. Thomas was administrator of the chemical demilitarization program for the Oregon Department of Environmental Quality.

Michael O'Hanlon is a Senior Fellow at the Brookings Institution and he formerly worked at the Congressional Budget Office. He specializes in defense issues, leads the work on Brookings' Iraq Index, and has served as team leader on two Brookings studies on homeland security in the last 3 years. The latest Brookings study on homeland security is expected to be published early in 2006. Dr. O'Hanlon received a Ph.D. in public and international affairs from Princeton University. He is also a visiting lecturer at the Woodrow Wilson School at Princeton University.

As I said, Senator Feinstein will introduce our two California guests today.

One final note about the witnesses at today's hearing. I want to point out that I invited officials from both the Department of Homeland Security and the Department of Defense to testify. A witness from DOD was prepared to testify, but DHS, the agency charged with leading the response in the wake of an attack, informed me that FEMA was too busy at this time to send a witness and that no other witness could be made available. DOD was not inclined to send a witness if DHS witnesses were not going to testify. I find this regrettable, but look forward to hearing from both DHS and DOD in the future.

The United States must be prepared to respond to terrorist attacks. Hurricane Katrina exposed the weaknesses in our Nation's emergency preparedness. We must determine whether similar problems could occur with a terrorist attack.

I would like to thank Senator Feinstein, as usual, for helping me prepare for and plan for this hearing. She and I see eye to eye on matters of national security and the need to respond to terrorist attacks and it is always a privilege to work with her in a very bipartisan way on this problem that, after all, confronts all Americans equally.

Senator Feinstein, thank you. The floor is yours.

STATEMENT OF HON. DIANNE FEINSTEIN, A U.S. SENATOR FROM THE STATE OF CALIFORNIA

Senator FEINSTEIN. Thank you, Mr. Chairman. I appreciate the comments and I respond to them in kind. I thank you very much for holding these hearings. I thank our witnesses. It is certainly great to see Slade Gorton again. Welcome back. I will make my remarks very brief.

September 11 is now 4 years away, and since that time, it has been commonplace to say everything has changed since September 11. And one thing that was supposed to change was our readiness to respond to a catastrophe. With that in mind, the Department of Homeland Security was forged and a large number of departments were put together with a total of some 22,000 employees. It was supposed to be strengthened. The ability to plan was supposed to have been made greater.

I increasingly believe that at least with respect to emergency preparedness, September 11 did not change everything. I think Hurricane Katrina is testimony to that. With Hurricane Katrina, it was 5 days' warning, and yet there was not the ability to evacuate and there certainly was not the ability to respond adequately. That lack of response was inadequate on all levels, local, State, as well as Federal.

So today, we examine the question, are we adequately prepared? This isn't an academic debate. This is what could happen in the wake of a terrorist attack or a huge natural disaster on one of our cities. If the government response to Katrina is any indication, it would be a time of chaos and confusion with American life at risk.

I was pleased to see from Senator Gorton's prepared testimony that he and the 9/11 Public Discourse Project have strongly endorsed efforts to require that scarce homeland security resources be allocated based on the best possible risk analysis. This is an assessment of threat, vulnerability, and consequence, and as you know, Senator, we haven't achieved that yet because everyone wants their part of the homeland security pie regardless of whether the assessment of threat, vulnerability, and consequence indicates that they should have part of that pie.

I am the original cosponsor, along with Senator Cornyn, of Senate legislation to accomplish this. This legislation was not approved by the Senate. The companion legislation passed the House overwhelmingly and is now being considered as part of the USA PATRIOT Reauthorization Act conference.

I also hope that our witnesses will address the questions that I think most Americans share. Is the level of preparedness for catastrophe higher than it was in 2001? Have things, in fact, improved? If so, how did things go so horribly wrong in Louisiana and what needs to be done to make us safer?

It is my pleasure to introduce our two distinguished panelists from my State, the State of California, Henry Renteria, Director of the State's Office of Emergency Services, and Matthew Bettenhausen, Director of the Governor's Office of Homeland Security. Respectively, they are charged with the operations and policies of the State of California in responding to disaster.

Over the course of 19 years as the head of the city of Oakland's Office of Emergency Services, Henry Renteria coordinated Oakland's response to eight Federal emergencies, including the Loma Prieta earthquake of 1989. That is the one in which a section of the Oakland Bay Bridge came down. More recently, he led the recovery efforts in San Joaquin County when a levee collapsed, flooding thousands of acres of farmland.

Matt Bettenhausen is a former Deputy Governor of Illinois with extensive law enforcement experience as a Federal prosecutor. He played a critical role in the development of the Department of Homeland Security as the first Director of State and Territorial Coordination, establishing the procedures linking State and Federal homeland security efforts.

Thank you very much, Mr. Chairman. I think we are ready for our witnesses.

Chairman KYL. Thank you very much, Senator Feinstein.

We are ready to begin, and Slade Gorton, the floor is yours.

STATEMENT OF HON. SLADE GORTON, 9/11 PUBLIC DISCOURSE PROJECT, SEATTLE, WASHINGTON

Senator GORTON. Thank you, Mr. Chairman. Senator Feinstein has already given a part of my speech, which was particularly flattering, but I will begin at the beginning.

If terrorists strike again on American soil, it will be local emergency responders—police, fire fighters, and emergency medical technicians—who will be on the front lines. Local emergency preparedness is now a matter of national security. In addition, of course, while the Federal Government through FEMA is not generally a first responder, its utterly inadequate response to the needs of both victims and first responders at the time of Katrina calls for dramatic changes in its preparation for and response to both natural and terrorist-caused emergencies.

On 9/11, shortcomings in emergency communications hindered first responders and led to an unnecessary loss of lives, especially bad among fire fighters in the Twin Towers and between agencies responding to the World Trade Center site. As the heroic fire fighters in both towers climbed higher, their radio transmissions were disrupted by the many floors between them and their commanders.

Communications with their chiefs in the lobby became weaker and more sporadic. Because so many people were trying to speak at once, available channels were overwhelmed. Transmissions overlapped and often became indecipherable. Many fire fighters in the North Tower didn't hear the evacuation order issued after the South Tower collapsed. Some weren't even aware that it had come down.

Meanwhile, communications among agencies were poor. In one well-known example, fire chiefs in the lobbies of the tower got no information from the police helicopters circling above. Because of poor inter-agency communications, many redundant searches were conducted. This wasted precious time and caused the deaths of many heroic first responders.

Hurricane Katrina reminds us that this problem has not been solved. In Katrina, poor public safety communications again delayed the response. New Orleans and the three neighboring parishes were using different equipment and different frequencies. They couldn't talk to one another. Helicopter crews couldn't talk to rescuers in boats. National Guard commanders in Mississippi had to use human couriers to carry messages.

Last July, the 9/11 Commission recommended that Congress turn over broadcast spectrum to first responders to improve communication between agencies and to allow interoperability among agencies. The House and Senate are finally moving on legislation to reclaim analog TV spectrum currently held by broadcasters and to designate some of it for use by emergency responders, but the date in the bill just released by the Commerce Committee is April 7, 2009, nearly 8 years after the 9/11 attacks.

By contrast, less than 4 years after Pearl Harbor, both Japan and Germany had been defeated. It is ridiculous that it should take eight years to implement such an obvious response to the 9/11 attacks. Experts say that this transition could be accomplished as early as 18 months from today and certainly within 2 years. There will surely be another terrorist attack or a major disaster in the next 4 years. We need a sense of urgency to get this done now, not 4 years from now.

On 9/11 in New York and New Orleans, command structures for emergency response were not clearly defined. It was not clear beforehand who was in charge or what each agency's responsibilities were. This confusion cost lives.

By contrast, in Arlington, Virginia at the Pentagon, the command structure did work and there was not loss of lives among first responders after the attack took place. I also have the impression that Mississippi's response to Katrina did not suffer from the same problems of command and control as did that of Louisiana. Command and control in response to Hurricane Rita seems to have worked better, as well. The Committee may well wish to examine the facts and circumstances of command and control in each of these cases so that we can learn what worked and what didn't.

The 9/11 Commission recommended that local governments adopt the incident command system. This system defines who is in charge and what agencies' responsibilities are in a crisis. Every locality should have a clear emergency plan with every agency's specific role laid out beforehand in black and white. As we saw in

Katrina, if local plans are not highly specific and are not regularly rehearsed, confusion is inevitable.

DHS set a hard deadline of October 1, 2006, for localities to establish and exercise a command and control system to qualify for first responder grants. Don't let that deadline slip. Localities that do not have clear, well-rehearsed incident command plans by that date should not receive Federal Homeland Security grants.

Now, Senator Feinstein has already spoken to risk-based funding. Since 2001, you have allocated more than $8 billion to help State and local governments prepare for terrorist attacks. Unfortunately, these funds have not been guided by any assessment of risk and vulnerability. The 9/11 Commission made a common sense recommendation that it be based on risks and vulnerabilities, not politics. These funds are national security funds, they are not general revenue sharing. They are too important to be spent without any guarantee that they are actually reducing our vulnerabilities.

Both of you support this kind of reform, as do many other Senators. As Senator Feinstein said, the House proposal on the subject is an excellent one. It is now in conference with the Senate. The Chairman and Vice Chairman of the 9/11 Commission and the Public Discourse Project will soon submit a letter to that Conference Committee strongly recommending the House version, which is very similar to what Senators Cornyn and Feinstein proposed here in the Senate. We urge that that be adopted.

The Intelligence Reform Act required DHS to produce a national strategy for transportation security by April 1, 2005. We bitterly criticized the agency for not having done so in our report in mid-September. We were told the next day, well, in fact, they had finished it on the first of July, but that it was classified, apparently, even the fact that it existed. As such, it is unavailable to the public, the transportation community, State and local governments, and first responders. What use is it if the people who have to adapt to it don't know anything about its existence or what it says?

Next, DHS has not produced the National Risk and Vulnerabilities Assessment for critical infrastructures that was due on June 15, making it very, very difficult to distribute Homeland Security funds in a rational manner. Moreover, that kind of assessment needs to be an ongoing process. It is not a one-time job.

Finally, as Hurricane Katrina reminded us, large-scale emergency responses are bound to occur again in the future, whether from terrorist attacks or natural disasters. The question, Mr. Chairman, is are we better prepared for the next major terrorist attack, for the next natural disaster? Are we prepared for an attack with a dirty bomb or one with chemical or biological weapons? Are our emergency communications good enough? Are our response plans updated and rehearsed? Are we directing Federal funds as they are needed to protect our real vulnerabilities?

Well, at least with respect to those last questions, the answer is no. After 9/11, after Katrina, we are still not prepared. We will do anything we can to help you and your counterparts in this Committee and in both Houses to enact these common-sense recommendations this year for the safety of our first responders and the communities they are pledged to protect. The lessons of 9/11 and again of Katrina are too painful to be learned again a third time.

[The prepared statement of Senator Gorton appears as a submission for the record.]

Chairman KYL. Thank you, Senator Gorton, for that excellent statement.

Next, Mr. Wayne Thomas.

STATEMENT OF WAYNE C. THOMAS, VICE PRESIDENT, HOMELAND SECURITY, INNOVATIVE EMERGENCY MANAGEMENT, INC., BATON ROUGE, LOUISIANA

Mr. THOMAS. Chairman Kyl, Ranking Member Senator Feinstein, other members of the Committee, I appreciate the opportunity and the honor to be here today. It is very significant for our company to have this opportunity and we appreciate it.

On the screen, you will see an image of a dirty bomb attack that we projected on the West side of the Capitol building here. I think this brings home to us what we are dealing with as a potential scenario. Should this event actually occur, part of this area may be uninhabitable for many, many years because it is a radiological device.

Recently, we witnessed the devastation of Hurricane Katrina and what happened along the Gulf Coast. I think we want to look at a comparison between terrorist events and the natural disasters that we just experienced.

The image that you are seeing now shows the extent of the impact of Hurricane Katrina and this model that you are seeing here is the storm surge model. We actually ran this model a year ago for an exercise that we did and we were aware at that time that this would be the extent of damage from a hurricane of this type. So we are projecting the geographical extent of the damage here into an image. We are going to compare this with a couple of other types of scenarios.

The first one that you will see is an IED event. This happens to be in Salt Lake City at the University of Utah stadium. This is a vehicle-mounted IED, and you will see that there are multiple IED attacks that happen in this one scenario that we have utilized. The message here is that this type of an attack is somewhat localized to the area, that the damage and the casualties would be localized, but the damage and effect on certainly the population where this happens in the Nation would be significant.

Second, you are going to see a chemical warfare attack in San Francisco at Golden Gate Park. This is a release of GB agent, Sarin, the same chemical agent that was used in Tokyo in 1996. A very small quantity is used here. The effects on those in the park would obviously be catastrophic, and we will project at the end of the simulation the casualties that we are looking at.

The last scenario that we have here is an anthrax attack on a Midwest city. I think once you see the simulation, you will see what city it is. This is a simulation that we did a couple of years ago and it is an airborne release of weaponized anthrax that impacts the entire city and beyond. In the bottom right corner, you can see the red and yellow images appearing. That is the distribution and spreading of the anthrax as it migrates through the communities. This would be absolutely catastrophic in this location or any urban environment were this to occur.

So how do these events all compare? You can see that here is the estimated fatalities that we have from these simulations. You can see that Hurricane Katrina that we saw and continue to see in New Orleans is less than we would anticipate from the bioterrorism anthrax simulation.

Returning back to the dirty bomb scenario, I think it is important that we also think about the Hurricane Wilma that recently crossed Florida and caused substantial damage there, went right up the East Coast very quickly and a few days ago was causing quite a bit of rain here. If that had moved a little bit to the West, it would have impacted D.C. The combination of a terrorist event along with a natural disaster could also be a very significant event.

I want to mention a few points here that from our experience working with many of our clients across the Nation and over the last 20 years, what we have seen from our experience. There are, in terms of catastrophic planning, I think various things that have been done, and one of the things that we have done back in July of 2004 was to develop an exercise, a planning exercise approach that we called Hurricane Pam. This was a Category 3 hurricane that made landfall in Southeast Louisiana. The consequences of that were used by the local communities, the State, and the Federal agencies for planning and 14 plans were developed at that time.

The consequences of a Hurricane Katrina-type event were well known in that area and by all of the response agencies. In fact, the model that we showed earlier demonstrated the extent of the storm surge that we knew well over a year ago would happen should this storm make landfall in this particular area.

The key aspect that we see working with our clients is that planning is the cornerstone of really everything that we do. But what we don't see is that we don't define what we want as specific, acceptable results from developing these plans. What is it we want to achieve? What is it that the public demands of us?

We write plans, we execute those plans, but we don't always define what we want, and that is a very important distinction that we need to address. Simply having a plan that works well may not achieve the results that you want if you don't determine what you want to achieve in advance, and that is a very significant change in planning approach that we would recommend.

The second issue is the actual plans that we developed, do we really understand the consequences that these disasters are going to have on our communities and on our citizens? We need that comprehensive understanding of a terrorist attack, whatever type we want to consider, natural hazards, hurricanes, earthquakes. Let us understand what the consequences are, because that helps us plan effectively. Unless we utilize those detailed consequence assessments, again, we cannot plan effectively.

The third point I would make is that we have to address the full integration of Federal, State, and local response capabilities. As Senator Gorton said, local response is first-line national response now and that is so accurate. All politics is local. I think Tip O'Neill said that many years ago. But all disasters are local. It is those local first responders that are going to be there when it immediately happens and they are going to be there over the years for the clean-up and recovery from that disaster. They are critical to solving the solutions here. So bringing the right people together is very important from the front-line local governments, State governments, and Federal agencies.

And the last thing I wanted to mention is that we have an exercise program that we utilize to test our plans. I think it is important that we make that program as rigorous as it can be. We conduct a lot of exercises that are essentially open book. We just test the plan and we check the marks here. But we have to have an exercise program that is rigorous, that ensures that we can effectively do what we say we think we can do.

I will sum it up in one statement, if I can. We have to plan together, train together, exercise together, and that puts us where we can respond and recover together as a nation. Thank you, sir.

[The prepared statement of Mr. Thomas appears as a submission for the record.]

Chairman KYL. Thank you, Mr. Thomas.

Mr. Renteria?

STATEMENT OF HENRY R. RENTERIA, DIRECTOR, CALIFORNIA GOVERNOR'S OFFICE OF EMERGENCY SERVICES, MATHER, CALIFORNIA

Mr. RENTERIA. Thank you and good morning, Chairman Kyl, Ranking Member Feinstein, and Subcommittee members. Thank you for the opportunity you have given me to be here today to testify on this very important subject.

Before I go into my remarks, I would like to, if we can, show a very brief video that we have brought along with us.

[A videotape was played.]

Mr. RENTERIA. Thank you. As you can see from the video, California, because of our history, and we do have a long history of natural disasters, we learned very many valuable lessons. We have taken these lessons and we have incorporated them into planning for the next one. In summary, we have had our Katrinas in California. We have had several disasters. We have learned from every single one of them and we have applied them to the next one.

As you saw, the Incident Command System is something that came out of the Forest Service. This was something that was used to manage large numbers of resources fighting large forest fires. This Incident Command System has now been pretty much developed into the civilian system as our Standardized Emergency Management System. SEMS, as we like to refer to it, is our bible for responding to disasters. This is the backbone of the system that is in California. Because this is an organizational system that can be used to deal with large-scale events, it allows the opportunity for several agencies to respond together under a unified command and deal with the events of the disaster.

As you saw from the video also, the Federal Government has now adopted the National Incident Management System, NIMS. It is a real tribute to California that they took our system and applied it, but I must also point out, the Federal Government did not use NIMS at the Katrina event and this was really a major issue that led to the things going wrong. If they had used NIMS as we have depicted here, I think we wouldn't have seen some of the issues that came up.

I also want to use this opportunity also to talk a little bit about interoperability. We have heard that word before and that is a major concern that we have. But I also want to show you that interoperability is something that we have taken very seriously, but it is also exacerbated by the fact that we are such a large State. We have unique topography in California that gives us major challenges and we have a huge number of response agencies.

But we have some success stories. I think this is something that I would like to point out. We have purchased what we call our black boxes. These are pieces of equipment that can be brought to a scene where first responders can literally plug into these boxes in order to solve some of the interoperability problems. We also have radio caches that we bring out to the scene to distribute to first responders who are showing up that may not have the radio frequencies that we have.

We have identified some success stories on a regional level, specifically in San Diego, Orange County, Sacramento, and even the Bay area. They have used some of the resources we have gotten from Homeland Security to develop some regional capabilities that gives us a model to follow for the rest of the State.

But we still need help. We need some guidance. The Federal Government needs to provide us some guidance on interoperability. What are those standards that you want us to follow? And we need the help with the frequencies, as was mentioned earlier. Frequencies is a major issue for us all over the country, and these frequencies and the spectrum that we need to have addressed so that we can have our interoperability taken care of.

I also want to point out, someone asked, what is the proper role of the Federal Government in a disaster? My response is, the Federal Government needs to be a partner. They need to be a partner with the State and with local government before, during, and after a disaster. Before the disaster, we must all speak with a single voice in helping spread the message of preparedness, preparedness from the level of the government, preparedness at the private sector, and preparedness for the individual citizens.

During a disaster, they need to also bring the resources in to help us respond and save lives, protect property. But also during a disaster, the Federal Government must also be part of a unified command. We have a system set up in California. We incorporate the Federal Government when they respond to our request, and so part of that unified command must be there for them to also participate in.

And after a disaster, besides bringing disaster assistance, which we obviously need, the Federal Government also needs to help us promote mitigation. The mitigation programs are the key to preventing some of the loss of life and property that we have in some of our natural disasters. We need to spend some money ahead of time so that we don't spend so much money after the event.

As we go forward and we identify the lessons learned, not only from Katrina but from past disasters, we will keep applying these lessons to our plans, we will keep exercising these plans, and we will keep providing more training to the necessary governments that need to be prepared to respond. But we must also remember the old saying that failure is only the opportunity to begin again intelligently. I think these are the lessons that we need to prepare for. We need to prepare for the next disaster, not the last one, and I think we are on that road. Thank you very much.

[The prepared statement of Mr. Renteria appears as a submission for the record.]

Chairman KYL. Thank you, Mr. Renteria.

Mr. Bettenhausen?

STATEMENT OF MATTHEW R. BETTENHAUSEN, DIRECTOR, CALIFORNIA OFFICE OF HOMELAND SECURITY, SACRAMENTO, CALIFORNIA

Mr. BETTENHAUSEN. I want to thank the Committee and each of the members for the opportunity to be here and also to praise your leadership on these important issues of homeland security. I know, Senator Kyl, we have many shared issues on border security in California that you have and we appreciate your leadership in providing the additional resources that we need to better secure America's borders.

Senator Feinstein, you have been a great partner with Governor Schwarzenegger, and just as you operate here in Washington, D.C. in a bipartisan fashion, we know that you have been a great partner with the Governor, that we work on homeland security and emergency management issues in a nonpartisan, bipartisan way, and we appreciate your support.

It has been a pleasure for me to join the Governor's team, and he has assembled a great team, including my partner here, Henry Renteria, but across the board, from food and agriculture to health services and we are working together and we appreciate your support and leadership here, and your work with Senator Cornyn, whom I have also had the pleasure of working with your daughter, Danley, when I was with the Department of Homeland Security.

You have been great leaders in recognizing that our funding must be prioritized, that it must be based on risk, that we must look at threat, vulnerability, and consequence. We have limited Federal resources and we must prioritize those and I thank you both and this Committee for its leadership.

I also would like to recognize Senator Durbin from my original home State of Illinois, where I also had the pleasure of being Deputy Governor and serving as its first Homeland Security Director and thank him for his leadership on homeland security issues, not only while I was there and leading the Illinois delegation, but working to continue making sure that Illinois, like California, is well prepared and continues to become better prepared as we look at these issues.

I think it is very appropriate that we look at the issue of terrorism in terms of the lessons learned from Katrina, and that is the focus that you have brought here today, and it is very important that we look at it from that perspective, not only from domestic terrorism that we learned from Timothy McVeigh and the Oklahoma City bombings, but through 9/11 to the 1993 World Trade Center bombings which helped save lives because there was better evacuation planning and there were things that were done when 9/11 happened 4 years ago.

But when we look at Katrina, and every one of those lives was precious, just as it was with 9/11, but if you look at the consequences there, where we lost approximately 1,200 individuals, each deeply important, major disruptions in their lives, but if you compare it to terrorism and if you compare it to 9/11, we had three times the number of casualties when we look at a potential terrorist attack.

As Senator Kyl appropriately pointed out, these attacks are not going to come—and Senator Feinstein, you did, as well—are not going to come with 5 days of warning. They are conspiring against us. Their intentions are well known by the intelligence community. It is well known by this Committee. They are looking for mass casualties. They are looking for a more spectacular event than 9/11. That means that we must keep the focus on terrorism preparedness, where we could have a weapon of mass destruction, a chemical, biological, radiological, nuclear, that could cause significant deaths and casualties.

One of the things that unfortunately or fortunately, depending on how you can kind of look at it, we have not been tested with major long-term care for mass casualties. In 9/11 and New York City, we were looking that we might have long-term care and needs for burn victims. Unfortunately, it didn't happen, and fortunately, I think because of the lessons learned from the 1993 bombing, there were a lot less casualties because there were evacuation plans and things in place that also saved lives.

But we have to go and look at what will we do in a major weapon of mass destruction event? What will be our medical surge capacity? What will FEMA be able to do in terms of providing support to our local, State, and counties in terms of that medical surge capacity that we know just does not exist in all of our communities when you are talking about an incident, and as Secretary Chertoff has recognized, that they are planning for housing and feeding—must account for 500,000 or more—or more—and it is not only housing and feeding, it is the medical surge capacity that would come with that. It comes with the ability to also bring communications, which is another lesson learned from Katrina, again from 9/11, and as Senator Gorton talked about, we haven't gotten it yet.

In 1997, Congress promised our first responders that they would have new frequencies and that we would have the capacity to have spectrum so that we could pass video, data, and have dedicated radio frequencies for our first responders. They were to get that at the end of this year. That is not going to happen, and I know the House markup is going on, that they are looking at a deadline for those spectrums to become available in 2008 and the Senate is looking at April of 2009. We need to move quickly to do that because we have learned again and again it is about communications, cooperation, and coordination. It is communications, communications, communications. So we need to do that.

I think one of the things that we look at as we move the media off those bands and that we as a Federal Government go to talk about auctioning those bands, as the government should—we have huge Federal deficits and we need the resources—we were meeting with Representative Lungren when we were out here. One of the things Congress should also consider is when we go and auction those things and we receive that money, that we dedicate that funding to our first responders so that they can use them to improve their interoperability and their communication capabilities, and it would be a good way to help finance it and make sure that we are committed to providing those resources.

Besides communications, again, we need to know that FEMA is going to have their logistics systems put in place so that when we call on FEMA, that we are going to be able to get the materials that we have requested that they have assured us that they are there and that we are going to be able to use them.

Besides housing and feeding, there is also the financing of the displaced. I think Secretary Chertoff has talked about this, about the need to improve it, and he is working on that, but it may be things that the Hill could look at in terms of improving Stafford Act so it is fairer to all States and all communities in terms of those impacted by disasters.

I think it is important to recognize, as Secretary Chertoff has, that FEMA is not a first responder. It is the State and locals. What is FEMA? It is nothing more than our first responders who are out there who have staff and are trained to take the urban search and rescue teams, the swift water rescue teams, that become a national resource in an emergency. Eight of California's USAR teams were in the Gulf.

We do not need to build Federal capabilities, as Governor Perry talked about, a Maytag repairman sort of sitting there waiting to be called. We need to better support our State and locals so they have those first responders there who can respond to anything 24/7, respond to incidents that don't require Federal assistance, but to be there for national emergencies as the national asset. So we can deploy USAR teams, swift water rescue, disaster medical assistance teams, and again, by USAR, I mean urban search and rescues.

Again, like Mr. Renteria talked about, we believe in and we need the support of the Federal Government, but we do not need the Federal Government in charge. We need the Federal Government to come in and assist us under civilian authority and control.

I would like to just talk briefly about IA, an information analysis and threat awareness. We need to continue to make improvements at all levels of government. I think the Federal Government and our intelligence community also needs to know that they can learn a lot from our local police officers who are out there. There are a lot more of them there. We know lessons learned in terms of Timothy McVeigh. It was local folks that got it. Eric Rudolph, it was local police who finally captured him.

We met and had our identity theft conference when we were in California, Senator, and I thank you for taking a leadership on that. We know that they need financing and they do it through the criminal milieu and there are things that we work at a local level that we can help provide, but we also need to know the strategic threats that are out there and we need better information sharing.

And then in terms of infrastructure protection, it was raised by Senator Gorton, as well, we need to have that plan. We need to have what the national infrastructure planning is going to be. We need to have better coordination with us in our private sector community in terms of which assets we need to harden and protect, and Senator, we have worked closely on many of those issues that we see in California, but we know that we can't harden all critical infrastructure, but we need what is the strategic system-wide plan so that we look at systems rather than individual targets and how do we have the redundancies, the resiliency, and the quick recovery capability so no matter where you hit us, and if you hit us in multiple places, we can quickly recover.

I see that my time has expired. I feel like I am back in the Court of Appeals with the clock running. But I would be remiss if I didn't talk about the importance of the individual citizen and the individual citizen's role to be prepared. That was really driven home with Katrina, Rita, and Wilma. You need to have a family disaster plan. You need to help our first responders so that they can address those most in need, those who are injured, and so that they can focus their attention on restoring services. So to the extent that you can take care of yourself, that you can be on your own for 72 hours, you are helping everybody. You are helping America and you are helping yourself and citizen preparedness is a critical part of making America safer and better prepared.

I thank you for the opportunity to speak with you today.

[The prepared statement of Mr. Bettenhausen appears as a submission for the record.]

Chairman KYL. Thank you, Mr. Bettenhausen.

Dr. O'Hanlon?

STATEMENT OF MICHAEL O'HANLON, SENIOR FELLOW AND CO-HOLDER, SYDNEY STEIN CHAIR, FOREIGN POLICY STUDIES PROGRAM, BROOKINGS INSTITUTION, WASHINGTON, D.C.

Mr. O'HANLON. Thank you, Senator. It is an honor to be here before you as well as Senator Feinstein. I also want to thank the Texas and Illinois delegation for giving us a great World Series so far, I know happier for some than others—

[Laughter.]

Mr. O'HANLON. —but too late for all of us. If we can agree on one thing, the games go too late. I think some people on this panel actually watched. Hopefully, you can't tell from our testimony.

Anyway, I want to say a few brief things very consistent with the general theme of what we have already heard from my colleagues.

I want to first say, in terms of answering, Senator Feinstein, your question about how much progress have we made and where do we stand, while there are a lot of things we haven't done and a lot of things we have, I want to put a little bit of a conceptual framework in place and suggest that we have put our greatest successful effort so far in the area of prevention, and that is appropriate. I think we have done better with the PATRIOT Act and with, even before the PATRIOT Act or concurrently with it, breaking down some of the bureaucratic barriers. The 9/11 Commission has offered some ideas on how to do more of this. But even before they issued their report, we had made some of these changes. I give the Bush administration and the Congress a lot of credit for that.

I think the border integration at DHS, the border security integration has gone better than some of the other efforts there, although it requires continued pressure from those of you who have a special interest in the border problem to keep up the resources, because I don't think they are adequate yet. But I do think that direction has been appropriate and the integration there has made good sense.

On a few other areas, maybe New York City's efforts to create dedicated police counterterror units and so forth, there have been some remarkable efforts at prevention, which I think have been helpful and it is part of why we haven't seen a lot of terrorists on American soil. We did learn from President Bush earlier this month in his October 6 speech that there had been some attacks in the works we hadn't known about, or at least those of us on the outside of government hadn't heard about prior to that point, but overall, I have been relieved that there haven't been more terrorists found on American soil since 9/11 and more attacks that actually got to the advanced planning stage or the implementation stage, and that is a tribute to our prevention and that should remain the top priority, I believe.

However, it cannot be the only priority because we are not good enough and our borders are inherently too open to make sure prevention always works. So we have to think a lot about protection of key assets and consequence management, and that is a lot of what my colleagues have been talking about, fellow panelists up here. I want to make a couple of observations before I get to my own graphics. Maybe I will conclude with those, but let me lay out a couple of broad thoughts first.

I would just offer my main recommendations on what we should do and what we shouldn't do at this moment in homeland security and next efforts, because I think we always have to be asking, what are our resources? What threats are most plausible that we should be preparing against them? And which ones are simply not plausible or too hard to deal with? Frankly, that latter category is a distressing one to have to recognize, but there are certain threats that unless we are prepared to do a radical change in our way of life or unless we see the threat get a lot more plausible, I think the best course of action is a fairly minimal response. In other areas, I think we need to do more. Let me just offer a couple of short lists of each, dos and don'ts.

On the don'ts, I don't believe, for example, that we should create a lot of excess hospital capacity for a quarantine in the event of a massive contagious biological attack. Some people have laid out very worrisome scenarios about contagious biological attack and I don't want to say these are implausible. In fact, we should spend a lot of time and effort on vaccines, on prevention, on monitoring people as they come into the country for health, trying to deal with health problems over seas. The H5N1 virus is an example of something we have to monitor and deal with through a health prevention approach.

But I don't think that we should spend what would be tens of billions of dollars creating excess hospital bed capacity for a scenario that is relatively unlikely to happen, and if we do have that scenario, it is going to be more important to respond in other ways than by having hospital beds. I throw out this scenario because some people have talked about the desirability of having the ability to have many, many thousands of people quarantined in any given city within hospitals in excess of what we already have—not a smart use of money, I don't believe.

There was a council on Foreign Relations study a couple of years ago that called for spending $20 billion a year more on first responders. I disagree. I don't think that would be a smart use of scarce homeland security money. We obviously have a lot of first responder needs that have not yet been met and I think there is room for an ample serious discussion on that topic alone, and we have heard some mention of initiatives that would be appropriate. I have my own list. But I think we have to keep that kind of list more or less within the $5 to $10 billion a year range that we have been spending so far on first responders because I don't think that some of the ideas that are out there make sense. Putting chemical protective suits for all three million first responders in the country at the top level of capacity, I don't think it is the appropriate thing to do.

Making sure every police and fire radio in the country can talk to every other one, I think that is excessive. I think what you need is mobile communications systems that can be interoperable, deployable ways for the fire and police radios to talk to each other. But to replace all the radios would be an excess use of resources.

I am sorry to go through this list of don'ts, but I want to establish some credibility, I hope, before I go to a list of dos, because we can really have a problem with homeland security of a kitchen sink mentality where those of us, most people in this room, I think, who are homeland security hawks sometimes sound to the rest of the country like we just want to do every single thing we can possibly imagine, not that anybody has been guilty of that here, but sometimes the impression people get is that homeland security hawks just want to spend everything under the sun, throw in the kitchen sink, at this problem. We have to avoid that temptation.

Another potential way you could spend umpteen sums of money would be to essentially harden our public spaces the way Israel has had to do. I do not believe we are now at a point in the United States where every single mall, restaurant, McDonald's, movie theater should have metal detectors. We may wind up in that world, and the Israelis have wound up in that world. I don't think we should be in that world right now because, again, I think the cost would be excessive. The threat is not yet credible enough to me to advocate that. Now, I could be proven wrong tomorrow, but I don't—and we obviously need certain kinds of buildings to be protected in these sorts of ways and we always have to have the debate about which ones. But I think to establish an Israeli-level security system for every public space would be excessive.

Finally, I don't think we need to inspect every single container coming into the United States. That would be roughly a 20fold increase in capacity compared to what we do today. It would require major redesign of every major port in the country. It would require additional expense on a magnitude of maybe ten times what we spend now. I would not recommend that.

Having said those things I would not do, let me say four things I would do very quickly, show my pictures, and be done.

One, chemical plants. Chemical plants, at least the top couple thousand most dangerous chemical plants in the country are just not well enough protected today, and I will add one little point of commentary on Senator Gorton's chart where he called on the private sector to take primary responsibility here. I agree with him partially. The private sector must do a lot of this, but I think they need a nudge from Washington, because if you are an individual owner of a chemical plant, of course, you are not going to volunteer to be the first one to protect your plant better than standards require. All the economic incentives are against doing so, and why would you want to draw attention to yourself or admit that you might have a vulnerability?

So you may do a few things quietly, and some chemical plant owners have, but most have not and I think it would be unrealistic for Washington to expect them to. On the other hand, we can't mandate with the heavy hand of government that every chemical plant in the country hire 1,000 more security guards tomorrow. That would be one of those don'ts that we should not do. So we need to figure out some compromise, and I think Congress needs to look at this in more detail than it has so far.

Border capacity. I think the efforts of this Committee and others have been instrumental and exemplary, but still insufficient, and you know more about that than I, so I won't go on.

Local police capacity, and I am fascinated to hear what my colleagues from California have to say on this. I have some friends on the L.A. City Council and elsewhere who have been distressed that in Los Angeles, and I have heard similar stories about Chicago and St. Louis and Houston, other places, there is really not much dedicated capacity at the level of local police to do what New York City is doing, which is to try to—and New York City does remarkable things. They will send a police officer to a convention of mosquito spraying equipment to figure out if anybody is there who doesn't seem to belong and might want that equipment to spray anthrax.

I don't think most other police departments in the country have thought about how to do that sort of thing, or which buildings might be most vulnerable to truck bombs, and therefore, perhaps, they should not have parking garages beneath them, or if they do, there should be much more rigorous inspection. Now, we all know there are a couple of big buildings in major cities that have taken these sorts of precautions, but I think New York City is the only city that does this systematically at the level of capacity that is appropriate. So helping cities create more capacity for preventive efforts at the level of police, I think is an appropriate third priority after chemical plants and border capacity.

Last thing, and Katrina brings this to mind, we need to avoid a big polarized debate about DOD's future role in disaster response. Some people want to say the States should always be the first responders. As Mr. Renteria said, all disasters are local—I guess it was Mr. Thomas—all disasters are local, all disaster response has to be local. At some level, of course, that is certainly true, but there are emergencies for which DOD is the only plausible way to marshal the kind of capacity we need and DOD is not yet good enough at reacting urgently.

Historically, DOD has acted over a period of days. They are not yet good enough at acting within hours. They should be able to be. They don't need a lot of new units. They don't need a lot of new capacity. They may or may not even need a new exemption to posse comitatus, although I would advocate one myself, but what they do need is better planning to figure out how to deploy a lot of capacity quickly.

OK, so those are my dos and don'ts. My apologies for going on a little long. I want to very quickly go through a couple of graphics that are not quite at the level of professionalism of my colleague, so I will be quick, but thankfully, Senator, your staff helped me make them better than they would have been otherwise.

This is what an anthrax attack could do with an airborne dispenser, an airplane, cruise missile, crop duster, what have you, in Washington. The shaded area is an area of high lethality and this would not require any more anthrax than you could have on one small airplane. So you are talking about potential for obviously many, many thousands, actually tens of thousands of deaths from this sort of an attack.

This is, of course, the worst case scenario, a hydrogen bomb. It is not a particularly likely terrorist threat. On the other hand, Russia still does have a lot of loose nukes that are man transportable, or certainly car transportable, and I don't think we have yet reached the point where we can feel good about the security of Russian nuclear materials. Graham Allison at Harvard was right, I think, to say we should have a Fort Knox standard for all plutonium and highly enriched uranium in the world. We should guard that as well as we guard gold, all of it, and we haven't yet gotten there, which means this threat is still plausible, hopefully very unlikely, but it is plausible even though terrorists cannot plausibly themselves enrich uranium or make plutonium.

This is another version of a dirty bomb and this is perhaps a somewhat less likely one, but it would be far worse than the graphs that Mr. Thomas showed earlier because what we are talking about here is cobalt from, for example, a food irradiation plant. Just one rod of this cobalt could actually create enough—if dispersed explosively could create enough contamination to look sort of like Chernobyl in terms of its effects and leave much of Manhattan uninhabitable for decades. Actually, I was surprised to learn this when I worked through a little bit of the science myself, but this is the sort of thing that we have somewhat unguarded, or at least not well enough guarded in our country today. So when you find a specific threat that could have this kind of implication, I think you do need to take preventive measures.

And this, of course, I won't expect you to read, but this is just a summary of what we have learned, and some of this was done at DOD in their preparation for thinking through terrorist scenarios, but you have got a short list of 15 with the typical casualty numbers in the thousands, typical property damage numbers in the tens to hundreds of billions, and most of these are things we haven't yet done enough to prepare against, so I will just quickly summarize that busy table with that comment and thank you for your patience.

[The prepared statement of Mr. O'Hanlon appears as a submission for the record.]

Chairman KYL. On that cheery news, we will conclude our panel discussion.

[Laughter.]

Chairman KYL. Obviously, you probably have to have a little humor to approach these extraordinarily serious subjects or it can literally keep you awake at night, but our job is to try to be as candid with the American people as possible, to bring these problems to their attention, and to do everything we can, along with our colleagues and those working with us at different levels of government, to be as prepared as we can for what Senator Gorton articulated was the inevitable terrorist attack of the future.

Senator Cornyn is going to have to leave, I think, shortly, and so I will call upon him now either for an opening statement or if he has a question or so with the concurrence of my colleagues here.

STATEMENT OF HON. JOHN CORNYN, A U.S. SENATOR FROM THE STATE OF TEXAS

Senator CORNYN. Thank you, Mr. Chairman. I just have a couple of things I want to highlight.

First, I would like to ask unanimous consent that my longer statement be made part of the record.

[The prepared statement of Senator Cornyn appears as a submission for the record.]

Senator CORNYN. I was glad to hear, and I knew Senator Feinstein, since she has been such a champion of this cause, would speak about risk-based funding, and I am glad to hear Senator Gorton reinforce that. That is simply a common sense, necessary fight that we have to keep pursuing. Unfortunately, there seems to be a tendency in Congress to want to split up money on a revenue sharing basis rather than on a risk basis, but simply put, the risk is too high for us to give up that fight, so we are going to keep pushing on that.

I would say, coming from a State that was affected, of course, both by evacuees from Katrina and then hit in Rita, how badly we are still—what kind of bad shape we are in terms of interoperability of communications. I can't tell you how many mayors I talked to who said basically they were operating on the basis of their cell phone. One said, "Well, I will give you my satellite telephone number if you need to call me." But in other places, even in a big city like Houston, they did not seem to have distribution on the necessary basis of interoperable communications. I just wonder, and I think we need to do more than wonder, we need to find out where all the money that Congress has appropriated for this purpose has been spent, because it looks like it has not been spent as well as it should.

The third issue I would highlight is continuity of government. This has been something that I have been concerned about. If one of those airplanes hadn't been brought down in Pennsylvania and hit the Capitol, it could have decapitated the Federal Government's ability to respond by killing or disabling a sufficient number of Senators and Congressmen that we would not be able to respond.

While the House has attempted to deal with its ability to constitute itself by providing for emergency elections in a 49-day period, all you have to do is look at the period of time after 9/11 to see how a much more immediate response is required than 49 days. Can you imagine running for election after a huge national emergency and just the difficulties of that? they have also attempted to deal with their quorum requirements by saying five members of the House can constitute a quorum and literally pass legislation, elect a Speaker and others, which I think has some constitutional problems, to say the least. So I hope we will continue working on that.

Finally, let me just talk about evacuation and cyber security. In the evacuations leading up to Rita, we saw that an order of local officials to evacuate 1.2 million people, because of the so-called Katrina effect, coming on the heels of a much more devastating hurricane, resulted in the evacuation of 2.7 million people, with our highways looking like parking lots, which caused frustration, but fortunately, no lives were lost and it was really nothing more than an inconvenience. But I think, obviously, we need to look at our evacuation plans, and it is not within one State, but I think, literally, a regional evacuation. So we need to look at that. Of course, it struck me that if we were indeed talking about a terrorist attack as opposed to a natural disaster, we would likely have no warning and thus no opportunity to evacuate.

Finally, Mr. Chairman, let me just mention cyber security is a cause that I have been interested in and concerned about for a long time. Of course, our ability to function in this economy with the first responders to get information is dependent on computers and our ability to basically include what is largely held in private hands, whether it is through financial institutions, local governments, or otherwise. We simply need to do a better job of protecting our cyber systems against computer attacks which could literally bring them down, disabling our first responders, affecting a body blow to our economy by bringing down our financial institutions or any one of a number of other scenarios you can think of that would be damaging, if not to life and limb, then certainly to our economy. And we need to do a better job through statutes like FISMA and others to enhance cyber security efforts, and I know the Department of Homeland Security is working on that, but I certainly don't think we are where we need to be.

Thank you for the opportunity to highlight a few of these issues.

Chairman KYL. Thank you, Senator Cornyn.

Let me begin with some questions, and I think just to have a good conversation, I will try to kind of stick to the 5-minute round concept here. We will just go back and forth. But if it takes longer for an answer or we have to go longer, as I said, we will not be strict about that.

First let me say that I think this is a very good panel to start a discussion, and I would hope that we will have the time to convene—maybe even ask some of you to come back in the future, but to continue the discussion because it is clear that this is not just a one-time proposition, that we can perhaps today only scratch the surface to identify some approaches and some of the problems and identify areas that we want to engage in in the future.

For example, the notion of cyber security that Senator Cornyn just mentioned, we have been involved in that literally for about 9 or 10 years. I remember when we first got involved in it, and that is an area—and one other—that gets into a lot of classified material that I really want to begin my questioning with.

One of you criticized—I believe it was Senator Gorton—that transportation plan that has been classified. This raises an interesting question of the dynamic between that which you don't want terrorists to know, but that which all of the people involved in—all of the public officials need to know and to some extent the public needs to know, the difference between a Katrina, for example, needing to know what the routes out of New Orleans are, and how we might respond to a terrorist attack. And terrorists, we know, from Iraq have gotten very good at planning the secondary attack. In other words, they draw you all to a place and then they create the real problem or they know what your exit or egress routes are, and that is where they plant the IEDs and so on.

So, anybody, starting with you, Senator Gorton, want to make at least some preliminary comments about the dynamic between that which necessarily does need to remain classified in the terrorist context, because that is the focus of hearing today, versus getting information out in the public?

Senator GORTON. Well, obviously, much of the work of our intelligence agencies about potential threats, about individuals, is quite appropriately classified. But one of the other panelists here mentioned the lack of desirability of examining, you know, every single container that comes into the United States by sea. You know, personally I agree with that statement.

Nevertheless, an overall transportation plan by the Department of Transportation is going to have to deal with that issue. When should they be examined? Under what circumstances should they be examined?

The people who are going to do the work in the ports here and elsewhere are going to have to know, you know, what those rules are. They are not, by and large, going to be people who have security clearances. And the difficulty here in the United States, literally almost forever, is the ease with which information is classified, the temptation once it is classified not to share it, often even with other agencies and the like, and the extreme difficultly of getting it declassified. This is just a particular example.

Are there elements in an overall transportation plan that we should not broadcast to the world? I am sure there are. But the existence of the plan and what people who are in the private sector need to know about the plan in order to carry out its recommendations? Of course, they should not be classified.

Chairman KYL. So one of the first things our Committee should do is to try to focus on some general principles with respect to the classification material so that that which needs to be classified is not overly restricted in sharing of it with the people that have to react to it and use it if there is a terrorist attack.

Senator GORTON. Yes.

Chairman KYL. We will try to work on that.

Now, let me just quickly turn—I found this interesting, that most of the visual illustrations did not appear to me to postulate worst-case scenarios by any means. In fact, all of you used wind coming from the right direction rather than the wrong direction, as I—well, no, excuse me. Actually, there were two. Dr. O'Hanlon, in yours the wind was going to from the southeast to the northwest to carry the radiation all the way up through Manhattan. But, Wayne Thomas, your explosion on the Mall blew it out toward the Lincoln Memorial rather than toward all of the Government buildings within the Federal enclave.

I am not sure what my question here is, but I guess is it that clearly a clever terrorist, knowing that to disperse anthrax or the radiation from a radiological weapon or chemicals understands, appreciates the importance of wind direction, will take those calculations in mind. And we know that they are very clever and calculating people, so that we are likely to have the worst-case scenario where wind direction is important for the effectiveness of a terrorist attack.

Would that be a fairly logical assumption, Mr. Thomas?

Mr. THOMAS. Yes, sir, it is a very logical assumption.

Chairman KYL. So we could easily have turned the wind direction around from west to east in your scenario, exploded the radiological device at the foot of the west side of the Capitol, and had a fairly major disaster for the Congressional office buildings, the Supreme Court, the Library of Congress, and the Capitol itself.

Mr. THOMAS. We certainly could have done that in the scenario. I think we were considering when we did that one that there would be an event happening on the Mall on the weekend where you would have a large tourist population and members of the public who would be impacted also. And it also went in the direction of the White House.

Chairman KYL. Well, there you go. No good can come of this, is the bottom line of that, and I thank you for pointing that out. Just don't tell the terrorists.

Boy, you have got to have a sense of humor in this, I think, or it gets very depressing very, very quickly. It is such a serious proposition.

Let me try to continue this conversation by calling on Senator Feinstein next, and then I will move into another series of questions.

Senator FEINSTEIN. Thanks very much, Mr. Chairman, and thank you very much for your testimony.

I think one of the great lessons of Katrina was that every mayor all across the Nation is going to be saying, "Oh, my God, I don't have an evacuation plan. And if I have an evacuation plan, clearly it is not going to be adequate."

I was a mayor for 9 years. I put a lot of effort into emergency planning, and what I found is that you have to have it all written and all rehearsed so that when something happens, the response is automatic and fast. You cannot wait a day or 2 days to make a decision.

So you have to know, if you need to get off-duty emergency forces back into your city, how you do it; if you have to use buses to evacuate people, how you do it. And I think what Katrina demonstrated is that the poor in a city need help and that there has to be a special plan given high priority, whereas people who don't have the resources are able to know where to go to be evacuated quickly. And I suspect nobody has that in their plan.

You know, I had to grapple with if there were a major earthquake, what streets would you bring in earth-moving equipment. Where would you get the earth-moving equipment? What companies would you go to? What yards would they be stored in? Really technical things, where emergency beds, medical supplies could be billeted for long-term use. I think we are led—how could we used closed military bases now as points for evacuation people to go to.

So I think there are a lot of things that Katrina brought up, but what I wanted to ask you is, Mr. Renteria, you mentioned that California will soon have its State interoperability plan. When will that be?

Mr. RENTERIA. We are very pleased to report that by January 2006 we will have a strategic plan for interoperability. As I mentioned earlier, it is a complex issue for California because of our topography and just the number of agencies, but because we do have some regional success stories, again, like San Diego, Orange County, we are hoping to tackle it from a regional standpoint and move it out.

But the committees that—

Senator FEINSTEIN. Will it have standards?

Mr. RENTERIA. That is what we are hoping to have, some standards. But, again, we need some guidance also from the Federal Government on those—

Senator FEINSTEIN. Now, where interoperability has been successfully developed—and by this I mean the ability of EMS personnel, sheriff's personnel, police personnel, maybe 10, 25 different departments, and maybe three or four counties, to talk to each other in the event of an emergency. Where it has been done, how has it been done successfully?

Mr. RENTERIA. That is, again, the San Diego model that we—

Senator FEINSTEIN. Could you tell us how it has been done successfully?

Mr. RENTERIA. They brought in all the players together, identified the radio systems, identified the different vendors that have been used, which is another issue that has to be addressed, different proprietary issues relative to the vendors and the private sector, because interoperability is beyond just radios. It is the ability to talk to each other, the standards, the common language, very similar to what SEMS is all about.

Senator FEINSTEIN. Are they all on a single megahertz?

Mr. RENTERIA. I don't think they are on a single megahertz. What they have done is identified the different systems that they have, and I don't think there are enough channels to put everybody on the same radio system, which, again, goes back to some of the issues you mentioned earlier.

But they have identified how they can work at that regional level to communicate with each other. I really want to explore more and communicate more with you on how they have actually done that so that we can—

Senator FEINSTEIN. Well, let me ask you, are there any standards for this? Because one thing that became rather clear to me is that when all the systems went down in Louisiana, particularly in New Orleans, they had no way of communicating police officer to police officer. No way. No satellite connection, no independent system.

So it seems to me that there are some standards that need to be put out there that every jurisdiction knows what they must do to have an emergency interoperable radio system up and running.

Mr. RENTERIA. Absolutely. And I am glad you pointed that out because one of the advantages we do have in California is that we do have some of those systems in place, like our Operational Area Satellite Information System. We refer to it as "OASIS." This is a satellite system that is in use and can be utilized by local governments also, that the State provides the communications devices for. So we do have some of those things in place. Our challenge is to make sure that we can expand it statewide, and that is what is going to be the biggest challenge.

Senator FEINSTEIN. [Presiding.] We have a vote that I believe there are 10 minutes left in the vote, and they are now cutting off the vote. One time I missed a vote because I got there 60 seconds after it had been cutoff, so I do not want to have that happen again. So what I am going to do is recess. I think Senator Kyl already went to vote, but I have to as well. So I will recess the Committee, and we will be back very shortly. Thank you.

[Recess 11:51 a.m. to 12:08 p.m.]

Chairman KYL. [Presiding.] Let's reconvene the hearing. My apologies. We had a vote called, and I thought maybe Senator Feinstein and I could play tag team, but I understand that the votes are being cutoff right at the designated time, which is odd for the Senate. And, therefore, Senator Feinstein did not want to miss that vote, and I do not blame her.

Again, with only 20 minutes or so to go here, let me just again thank everybody for kind of writing the preamble to what I want to move forward with. And there are so many different questions, so in my remaining time, I am just going to try to set the stage for some future meetings.

In that regard, several of you made points that tied together, and let me kind of summarize it, and then try to get the response from each of you.

First of all, Mr. Bettenhausen raised what I noted as a question specifically to ask you, but I would like all of you to think about it, and that is, the differences, if any, in planning for a natural disaster versus a terrorist attack. You mentioned the difference, for example, in burn casualties that might be expected in a terrorist attack versus most natural disasters.

Dr. O'Hanlon really did us a service, I think, by forcing us to concentrate on things not to do, not because they are not good things, but because you have got scarce resources, and inevitably we don't have time or resources to do everything we want to do. And I would add a third thing, and I think some of you alluded to this, too. We are such a big and open country that even if we wanted to do some things, it would be impossible, for example, harden every shopping center or the like, and, therefore, to try to basically provide some triage in the planning. And one of you said one of the first questions is—I think this was Mr. Thomas. You said the first question you have to ask is: What do you want to achieve in developing this plan?

And it seems to me there are at least three things. One, prevention. And I agree, Dr. O'Hanlon, that prevention needs to be our first defense. Second, protection, which is to some extent prevention, but it assumes that maybe something has happened or is happening. And, third, responding and all that that means. And we need plans with respect to all three, and it is not just the Federal Government's job. For example, on prevention, I think you talked about the chemical plants, for example, and all of you in one way or another have talked about the need for citizens to think about things that they could do in their own lives and how they would respond as well.

So you have got: What do we want to achieve? Between prevention and protection and response, how do we calibrate those? What kind of resources do we have in order to prioritize specifically what we do? And in the context of both natural disasters and terrorist attacks.

If I could kind of frame the question that way, it is impossible for all of you to adequately respond to that question right now. But let me ask all of you to give it a shot, and then add anything you would like to add for our record, and then we may call you all back again, or if that is not convenient, in some way get your advice in the future.

With that sort of four-part context, would all of you just like to tell me what you think we need to know in getting ready for more of these hearings based upon what I have said? Senator Gorton?

Senator GORTON. Well, Mr. Chairman, I think the principal distinction between a terrorist attack and a natural disaster is that, by definition, a successful terrorist attack has not been predicted. We are trying to predict them, but when we can predict them, we can probably prevent them.

But the first responders, the emergency responders are going to be responding to a specific incident that has not been predicted. Training for the disaster is vitally important. Most natural disasters are going to have been predicted. I think as Senator Feinstein said, we had 5 days' notice of Katrina in many respects.

Now, in our part of the country, we still don't predict earthquakes very well, and an earthquake is likely to be more like the terrorist attack than the hurricane is. But I think that is, you know, the primary distinction. But the training for both, it seems to me, is vitally important, the kind of things we talked about: a proper command structure, the ability to communicate when most communication lines are down, all that.

If I may indulge with two former colleagues, I would like to put one other thought in your mind, and that has to do with inevitable tendencies of any kind of governmental agency to have rules and regulations that are perfectly appropriate when you have got plenty of time, but that interfere when you don't.

If you and I, Mr. Chairman, were to change States, neither of us could practice law immediately upon going to the other State. We would have to go through some kind of procedure, which is quite reasonable. But I am a member of an organization that meets on various civic events every Monday, and two and a half weeks ago our speaker was the head of a marvelous volunteer organization of physicians and health care professionals in Oregon and Washington that sends volunteers all over the world to respond to health care emergencies. They sent people to the Indian Ocean at the time of the tsunami.

The day after Katrina, they sent a crew to Baton Rouge, where they were promptly told they were not licensed to practice medicine in the State of Louisiana. And it took somewhere between 24 and 72 hours before they could use those skills in that place. I would tell you this. This was reported to us by the head of the agency. I can certify it.

The other one I cannot certify, but I think you might want to check on it. I have been told that a large number of highly professional emergency responders, firefighting officials and the like, immediately went from the Northeast down to try to help. They were stopped in Atlanta and told they had to undergo at least 24 hours of sexual harassment training before they could be sent on to do the things that they do, you know, by FEMA.

Sure, people ought to be licensed to practice medicine. Sure, people ought to have appropriate training and the like. But one of the great inhibitors, it seems to me, is just that kind of mentality that rules that are perfectly OK in non-emergency situations are highly damaging and restrictive in emergency situations. And we and you at the Federal Government, we have got to see some way or another that emergency responders can respond instantaneously and promptly to the emergency and are not restricted by inhibitory rules of this nature.

Chairman KYL. Thank you. Very enlightening.

Mr. Thomas?

Mr. THOMAS. Mr. Chairman, I think your initial question was the difference between terrorist events and natural hazards. To me there is a very simple distinction between the two, and it involves prevention. We can prevent terrorist attacks. We cannot prevent natural hazards. They will happen. We could watch Hurricane Katrina march across the Gulf. We knew it was coming. We could not prevent the hurricane, but we could do what we needed to do to protect our citizens and get them out of the way. We knew it was affect our infrastructure. But that was all something that we could anticipate.

With a terrorist attack, we don't know where or when it will happen or what it will be. But an aggressive prevention program is a major distinction between the two.

I think the other question that you posed was: What do we want to achieve? I think I had mentioned that in my earlier testimony. I think that raises the difficult questions. You know, what is the outcome that we want in implementing these plans, protecting the public? Is it 100 percent? Is it 95 percent?

We talk about evacuating large cities. We know that certain parts of the population will refuse to go. Is that acceptable to us?

These are questions that we have not really addressed. Senator Feinstein mentioned the poor. How do we deal with them in terms of ensuring their safety should a disaster happen? I think these have been largely grouped into other issues and not singled out for discussion and have not been addressed.

So there is a tremendous number of issues that we need to get into here. You are right that we need a more substantial dialog on this.

I think when we talk about results, it is not do we have a prevention plan, do we have a response plan, and so forth. Those are kind of the easy things to say. What does that prevention plan do? If we implement it, how is it successful? Is preventing all terrorist attacks the only measure that we can have 100 percent success? Is evacuating 100 percent of the population the only measure?

Those are the things that we have to define, and I think as we look at those questions, that is where we have to use technology and other capabilities to come up with, first of all, the questions we want to address and then what are the acceptable solutions that we as, I think, collectively citizens want to have as our measures.

Thank you.

Chairman KYL. Thank you. Would each of you also address as we go on—I have been seeing something here, and that is, all of you believe that we need better plans, even if plans do exist.

Mr. Renteria?

Mr. RENTERIA. Yes. Without repeating some of the things mentioned earlier, I think some of the things that are the same have to do with the consequences of an event and the recovery of an event. Consequences, you are going to have people killed, you are going to have people hurt, you are going to have the property damaged or destroyed. So natural disasters, terrorists, human cause, whatever, you are going to have those same consequences.

The responses to the consequences are basically the same, too. We train to deal with these types of events every day. We must respond to them adequately.

The differences: When you have a terrorist event—and Matt can probably give you more information on this—you have a crime scene. That involves not just local law enforcement agencies but Federal. And that does complicate matters sometimes because some of our locals are not used to receiving this type of involvement.

And then the other thing that makes it different—and this brings up my old social work background—is the psychological effects of a terrorist event versus a natural event. All sorts of psychological studies would tell you people understand these "acts of God" or something that is going to happen, may happen, and all of us will be affected by it. But a terrorist event brings a whole other level of fear and trepidation on the part of people that they cannot go on with their normal lives.

To answer your question what we hope to achieve, for me it is cooperation and being unified in our preparedness, response, and recovery. We all need to be on the same page.

Chairman KYL. Thank you very much.

Mr. Bettenhausen?

Mr. BETTENHAUSEN. Senator, let me followup with one of the questions you earlier had, which I think gets to some of your prevention questions, protection, and responding. And that was the issue of the classified transportation plan, and I want to echo Senator Gorton's comments on that, but reiterate something even more important.

There are not going to be enough dollars—there is already a backlog—to provide security clearances with everybody who has a potential need to know. There was the thought with intelligence reform that we were going to change the way we were doing business, that instead of writing the classified, we would write to release; that we would truly embrace our partners, the first responders, and providing them with information; and not just information about potential threats but strategic information.

We know that there is a lot of intent to do us harm, but what is that intent? What are their capabilities? What are you picking up in terms of what should we prioritize and be doing first with the scarce dollars we have? Is it chemical? Is it biological? Is it radiological? Is it nuclear? But help us with the strategic planning. Write to release. Change the way that we are doing business. Nobody needs to know sources, methods, and means in terms of how intelligence is collected. But, you know, by God, if a cyanide truck has been hijacked and is missing, not only does law enforcement need to know where to look for it, our first responders, our hospitals, our private sector folks need to be dusting off their plans on how they would deal with, you know, a cyanide situation.

So we will never have enough clearances with the turnover and the backlog that we have even with the Federal Government. We need to make sure that we are taking us full on partnership and that we are looking at this information that needs to be shared strategically and across the entire first responder community, not just law enforcement but also fire, our public works, and our important private sector partners.

Another problem if you have something with a classified transportation report, it is a problem that we had with the sort of—the inside-the-Beltway thinking, that we are the Feds, we know how to do it, and, you know, you out in the hinterlands don't have a clue. That was the first problem with NIMS version 1.0. They never asked the State and locals how to do it. They never asked California: How have you been doing emergency management so effectively? Jeb Bush, how have you been doing it? And they went and they hired a contractor, and they came up with a very nice system that was then presented sort of as a fair accompli to State and locals and said, "This ain't going to work."

When you go and you have your transportation plan, who owns those things? Who operates it? It may be that the Federal Government needs to have a strategic plan overall, but you need to be talking with the State and locals, the ones who are running and working these things from the very beginning, not that you have come and thought this and now isn't this great and present it to us—and, well, actually not present it to us, classify it.

The other problem with classification, those of us who sit—I have my letter clearances up to SCI. To some extent, just don't pass the buck and say, look, I have now informed you, but you cannot tell anybody else. That does me no good. All that does it pass the buck. You need to help me with the information and bring that together, because, again, as you have pointed out, prevention is the No. 1

goal and priority, if we can stop it before it happens. And that's what the difference is in the planning in terms of what we need to do versus all hazards. You cannot turn—we cannot force a tornado to change. We cannot stop an earthquake from happening. But we can interdict, we can deter terrorist activities with better infrastructure protection, with better strategic intelligence, with better planning in terms of trying to prevent something from happening. That helps with the planning activities.

And then in terms of sort of terrorists, I think some of it is that you have—the potential for the mass casualties and the overwhelming of systems and things that you talked about with cyber, that if you had a magnetic pulse in terms of what that is going to do with our cars that now have electronics and what it is going to do to our computer systems, how are we going to deal with these issues that become just very large scale very quickly.

So those are some of the differences, but it is still the basics. As my good partner Henry Renteria was talking about, to some extent when you are a first responder and you are going to a collapsed building, you don't much care why it collapsed. Was it because of an earthquake? Was it because of a tornado? Was it knocked down by a hurricane? Or was it knocked down by the criminal acts of evil men? Our first priority is saving lives and preventing collateral damage, the kind of—even our domestic terrorists know to do the one-two strike. That's what we saw on 9/11. You know, Washington, D.C., New York. New York, tower one, tower two. I mean, it's an established thing, so you need to be thinking about it, and you have got those potentials to prevent.

Exercises, very critical in terms of doing it, and with FEMA, we've got a statewide exercise program. That helps to develop better plans. Exercises are not about patting yourself on the back about what a great job we can do and how we can respond to it. It really is about testing your system, overloading it, push it to the point of failure so that you know how you can write those plans better, make them better, and respond better in a real instant. It is doing after-action reports. You know, unfortunately, we have more than our fair share of natural disasters in California. But the fortunate aspect out of it is we get a lot of lessons learned. But you cannot lose those things, and you need to be asking the questions: What went right and what went wrong? And how do you make sure the right things are incorporated and duplicated and the wrong things are pulled out of the system?

The last thing, I forgot to thank, when I was thanking, the great staff that you have: the two Steves who we look forward to working with them, because I think you are right, we are just scratching the surface here, and to the extent that we can help you and work with your wonderful staff, we will to help flesh out some more of these issues.

Chairman KYL. Thank you very much, and you are right about our great staff.

We are now a little pressed for time, so, Dr. O'Hanlon, would you—and then I am going to turn to Senator Feinstein and leave and have her conclude the hearing. But if you could—

Mr. O'HANLON. I will just say a few things, Senator. I am struck by how hard this problem is. I have done most of my work in my career in defense, which I think is easy compared—national security, military issues, easy compared to these questions, because these questions involve local, State, and Federal efforts, they involve private sector. There is such a different type of threat from one area to another. I think it is just an inherently very challenging field of study.

Let me give one example and stop. There is one major city in this country—I will not say which one—that tried to say, OK, which of our big buildings do we really have to protect, provide this local site protection for, having done all the prevention, having prepared some consequence management. And they went through and in their first iteration came up with a couple thousand sites. And they said, well, there is a logic that gets you to a couple thousand, but it is just too many to protect, so we have to try again.

So they went through it, the same exercise, and got down to a few hundred. And they said it is still too many, and they finally did a third round and got to a few dozen.

Well, how do we know they got the right answer? They did not really get the right answer. They were constantly trading things off, one against the other. They came up against a problem that was manageable in size and yet still ambitious in scope. That underscores for me the nature of the challenge here.

One more example, and I will stop. Skyscraper. How do you protect skyscrapers? What is the appropriate level of protection? Is it making sure a truck bomb cannot be—cannot get within 100 feet by closing off side streets? Is it making sure that air intakes are all at least two stories above street level so people cannot put anthrax in? Is it making sure you have security guards who are well trained at every entrance so that people cannot sneak explosives up and create an apartment bomb the way they were worried in 2002 in New York?

I think these are almost unanswerable questions, and the only way in which you can work toward pretty good answers is to have dedicated study ongoing from committees like this, from commissions like the 9/11 Commission, and from experts in the field.

So it is just a way of saying this problem is inherently very hard, probably harder than any other problem I have studied in public policy. And so, therefore, I thank you for the ongoing attention from this Committee.

Senator FEINSTEIN. [Presiding.] Well, you are very welcome, and thank you.

Let me just begin by putting in a statement from Senator Cornyn into the record.

As I listen to this, and having functioned as a mayor, it occurs to me that what the Department of Homeland Security might be doing is preparing a series of advisory standards that can go to local and State jurisdictions in a number of different areas, operability being one of them.

What should the standards be as you consider interoperability? And what are your options? What kind of equipment is available? How much does it cost? Let the local jurisdictions make their own decisions, but there are some technical advisory standards that are available, certainly standards for evacuation of an area.

Now, in my city this becomes particularly dangerous. The uniform forces all live outside the city, across the bridges. If bridges come down, how do we get the uniform forces back? That has to be thought of long before a major earthquake or a terrorist attack.

How do you evacuate the poor? How do you evacuate hospitals if you have to, nursing homes? I thought St. Rita's was just a terrible example of the absence of any kind of overall policy with respect to a nursing home where people watched water draw up to them, and then obviously drowned in it because they could not move. I mean, it was terrible.

Emergency manuals. How do you prepare an emergency manual? What should be in that manual? How should you rehearse that manual? What kind of synthetic scenario should you practice out there based on your own individual needs geographically, politically, across this country? Standards for a family disaster plan. As you gentlemen have said today, families need to have their own plans.

Well, I deal with this all the time. I am sure what I have is inadequate. It is put together helter-skelter, didn't pay much attention to it. I do not have a checklist, those kinds of things. If you store water, how often should you change that water storage? How often should you change batteries?

I mean, just technical things that could go out to all Americans to know how they need to protect themselves in the first 48 hours of a disaster. You know, 22,000 people int he Department of Homeland Security, and this—

Mr. BETTENHAUSEN. It is 180,000.

Senator FEINSTEIN. I beg your pardon?

Mr. BETTENHAUSEN. It is 180,000.

Senator FEINSTEIN. You are right. Excuse me.

Mr. BETTENHAUSEN. It is 22 agencies.

Senator FEINSTEIN. That is right, 180,000, and they cannot seem to do this kind of work. I find that inexcusable because, to me, it is a no-brainer of a way that, without much cost, the Federal Government could use its expertise, use its reach, and use its ability to bring people together to prepare something which could be of real practice use for local jurisdictions.

It may well be that the State of California can play a leadership role in this respect, and hopefully when your plan comes out in January you will share it with a lot of us so that we can take a good look at it and see what you have done.

I would appreciate—and I think Senator Kyl would as well—recommendations of what should we do with respect to the spectrum now. How should we proceed? Those recommendations I think would be very effective.

But I don't really have any other questions. If you have any closing comments in addition, I would be happy to hear them before I close the hearing.

Senator GORTON. Senator Feinstein, on that very last point, I believe Senator McCain will propose an amendment when that Commerce Committee bill comes to shorten the date on spectrum transfer.

Senator FEINSTEIN. Oh, good. I am happy to hear that.

Senator GORTON. It will be a tough vote, but I think he will have that opportunity.

Senator FEINSTEIN. Good.

Anybody else have a last comment? Mr. Renteria?

Mr. RENTERIA. Yes. First of all, I would like to thank Matt for helping us thank everybody, because I think he helped us remember all the names. But I also wanted to invite you and other Committee members to our exercise that we are going to have in California on November 15th. It is the Golden Guardian event. Matt Bettenhausen and I have been working very closely on it. His office is funding it. And so this is another opportunity for you to see—and all of your staff to see, also—what we are doing in California. We are going to include an interoperability component also, so you can see—

Senator FEINSTEIN. I would like to come. I will probably be in the Miers hearings at that time if they are going on, but I will certainly have my staff be there.

Mr. RENTERIA. I think we can have somebody show up.

Senator FEINSTEIN. I appreciate it.

Mr. RENTERIA. Thank you again for this opportunity.

Senator FEINSTEIN. You are very welcome.

Mr. Bettenhausen?

Mr. BETTENHAUSEN. And that is part of what another famous Californian said with exercise. We will also be working closely with FEMA, so we will trust but verify by doing the exercises as well.

But I think you also pointed out with our poor communities, what you knew as a mayor, and what we also know from the State standards. It is not just poor communities that we need to make sure that we are doing extra planning for. It is also our special needs citizens that are out there, and it is one of the things that we have seen in terms of the exercises and practices we do with our nuclear power plants in terms of identifying those folks.

But, you know, it is another role that citizens can plan, and we saw a lot of that with Katrina and Rita, knowing in your neighborhoods who needs help and helping us help identify them and helping—you know, being able to take care of your family and then also helping your neighbors.

Senator FEINSTEIN. Yes, I think even planning where to go. Most people, I would hazard a guess, don't know where they would go in an emergency. But having something that is realistic, that gets you out of the immediate area where you can sustain yourself for a period of time, making those arrangements ahead of time. And I do think that is the world that we live in, that we know there is some place we can go, where there is some help around us.

Mr. BETTENHAUSEN. Two things to followup, too. Not only does the California Office of Homeland Security and Emergency Management have a sort of—have a preparedness campaign, because the First Lady of California has taken this on as a personal mission, and she has been doing a great job.

Senator FEINSTEIN. That is great.

Mr. BETTENHAUSEN. The people who have turned out, to become—be responsible for yourselves and be prepared and be ready. But www.ready.gov has the listing of how you can prepare a personal communications, a family communications plan, the kind of kits that you have. And an easy way to remember how to sort of restock in and out of there is, as you turn your clocks forward and back, switch the canned goods and the water out of it. The same thing that you should be changing your smoke detector batteries. It is an easy way to remember it. That is the time, you know, when you are doing those to—

Senator FEINSTEIN. That is a very good idea.

Mr. BETTENHAUSEN. That is when you should be doing the rest of your work as well.

Senator FEINSTEIN. See, I think to some extent we get so esoteric, when most of this is good, solid, practical planning ahead of time. And I do not believe that the cities of America are really equipped for a major disaster.

Let me just end with one thing. I think it is very important. In California, we have one American city that doesn't have 100-year flood protection, and that is Sacramento. And the people of Sacramento should know this. If we have an earthquake and the levees go down, the flooding potential for Sacramento is enormous. And I very much appreciate the fact that the Governor wants to be of help. We want to try to get some money to facilitate maximum levee repair within a reasonably short period of time to protect the city against the loss of human life. And I think we have agreed that that is our No. 1 priority.

I met yesterday with Congresswoman Matsui. I know that is hers. And we would really welcome continuing working with the Governor's office to see that that happens.

Mr. BETTENHAUSEN. And the entire California delegation.

Senator FEINSTEIN. Yes, absolutely. Absolutely.

Mr. O'Hanlon, do you have a comment to end this thing on?

Mr. O'HANLON. No. Thank you, Senator.

Senator FEINSTEIN. Well, thank you all very, very much. The hearing is adjourned. Thank you.

[Whereupon, at 12:38 p.m., the Subcommittee was adjourned.]

[Questions and answers and submissions for the record follow.]

QUESTIONS AND ANSWERS

Questions for Michael O'Hanlon
From Senator Joseph R. Biden, Jr.
Terrorism Emergency Preparedness
Senate Subcommittee on Terrorism, Technology, and Homeland Security
October 26, 2005

Transport of Hazardous Materials by Rail

In your testimony you were asked to explore the likely the likely consequences of several potentially severe terrorist scenarios and to assess what steps may have been taken to assess the risk. Your testimony in this regard was quite extensive; however, one area that you did not explore was the likelihood and consequences of 90-ton rail tanker filled with toxic-by-inhalation chemicals, such as chlorine. As you know, thousands of these tankers are shipped over unguarded tracks through our metropolitan areas every day. There is very little fencing or security along many of these routes and journalists, red teams, and normal trespassers have easily accessed the tracks.

The Bush Administration has not, in my opinion, taken this problem seriously. They have failed to develop a comprehensive strategy for handling rail security, and they even joined the rail industry in a law suit to enjoin a District of Columbia law to re-route these shipments out of the city. Given the Federal government's lax approach, I understand the reasons the District of Columbia has taken this action and other cities such as Boston, Baltimore, and Chicago are considering similar action. I firmly believe that this problem requires a comprehensive national approach, and I have introduced Senate bill S.1256 that would require the Department of Homeland Security to develop a nation strategy to address the concerns related to the transport of hazardous chemicals over the rails. While I would not ask you to review S. 1256, I would like to ask you a few questions related to this issue:

Question: What in your opinion is the state of freight rail security in our nation?

Answer: Overall the state is poor. This is partly inevitable, given the difficulty of protecting freight rail, but also dangerous, given certain specific serious risks, with chlorine and other highly toxic chemicals topping the list of importance in my opinion. That which can be reasonably done to mitigate the latter types of risks should be done.

Question: What is your estimation of the damage (death, injury, and economic) if rail tanker carrying chlorine were successfully attacked in a metropolitan area. Please assume a worst case scenario with respect to temperature, wind direction, etc.

Answer: A "reasonable worst case" – that is, bad, but not hyped – could easily cause thousands of fatalities and quite possibly tens of thousands (or even more).

Question: In your opinion, is this an area where a comprehensive strategy is needed and can you briefly assess what the federal government has done in response.

Answer: Yes this is precisely the kind of "catastrophic threat" that we at Brookings have emphasized in our studies since 9/11, and it is one of the relatively few for which government's actions to date have been wholly inadequate.

Role of Local Police

While the issue of the role of local police in our homeland security efforts was not a part of your direct testimony, I would like for you to expand on your August 24, 2004 article discussing how local police can intervene. In this article, you state that "[c]ops on the beat are a necessary part of the answer. They know their neighborhoods and often have hunches about who may be up to no good. the provide community policing, track identity theft and marriage fraud, and develop trusted local sources. They are in the best position to 'collect' the dots that the federal agencies need to 'connect' to forecast the next attack." I couldn't agree more, but unfortunately, due to local budget cuts and drastic cuts to the Office of Community Oriented Policing Services (COPS) many local agencies have a drastic shortage of officers. Not only does this hinder their local crime and homeland security efforts, it also hinders their ability to participate in federal terrorism task forces. In your article, you mention that only 14 officers in the Los Angeles Police Department participate in the Joint Terrorism Task Force of the FBI. In addition, the transfer of FBI agents from local crime to anti-terror cases is placing further strain on local agencies, which is reducing participation even further. Based on my conversations with local officials, this is a nationwide problem.

Question: Do you believe that officer shortages at the local level, which limit the ability to engage in special programs such as community policing and participation in anti-terror efforts, ultimately harms our nation's homeland security and anti-terror efforts.

Answer: Absolutely, yes.

Question: Many of my colleagues have justified federal cuts to local law enforcement because they believe that local law enforcement is a local concern. I have always disagreed as I believe that public safety is the most important function of the federal government. This is especially true in the post 9/11 world when it is possible, if not probable, that terrorists are in our local communities making attack plans. In your opinion, does the important role of local law enforcement in our national anti-terror efforts impact this restricted view of roles and responsibilities?

Answer: This is a very important issue in American federalism. Of course philosophical views about the proper role of federal versus state and local government inevitably shape people's views. My view is that some calls to greatly expand federal support for policing under the guise of the homeland security mission can go too far, and I think some proposals for greatly increasing funding for first responders have been too lavish (such as the Council on Foreign Relations study a couple of years ago advocating $20 Bn/yr in additional funding). That said, I absolutely agree that security is a federal responsibility and that where specific homeland security needs are identified, the federal government should help. In this regard, the dearth of funding for creating additional police capacity for homeland security related tasks since 9/11 has been a national policy mistake in my judgment, and should be corrected.

SUBMISSIONS FOR THE RECORD

STATEMENT OF DIRECTOR MATTHEW R. BETTENHAUSEN CALIFORNIA OFFICE OF HOMELAND SECURITY BEFORE THE SENATE JUDICAIRY SUBCOMMITTEE ON TERRORISM, TECHNOLOGY AND HOMELAND SECURITY

October 26, 2005

Mr. Chairman and Members of the Committee: Thank you for the opportunity to address you today.

In the wake of the September 11 attacks, California, along with the rest of the nation moved quickly to respond to the threat of international terrorism. Al Qaeda was quickly identified as the perpetrator of the attacks and remains a primary focus of the ongoing terrorist threat to America. Their goal is to cause mass casualties and inflict damage upon America's economy and national icons. California continues to be a high risk state with a wide variety of potential targets.

The nature of this threat demands that measures be taken with a sense of urgency to address a broad variety of needs. Systems for a wide variety of homeland security needs have been put in place, such as information sharing and threat assessment, grants management, critical infrastructure protection, and training and exercises.

California, in partnership with federal and local governments and the private sector, has prioritized its efforts across the homeland security spectrum of: prevention, preparedness, response and recovery. In achieving these homeland security goals, California believes that it is vital to develop these capabilities across disciplines and on a regional basis. California's Homeland Security Strategy promotes information sharing as detailed in the National Homeland Security Strategy and the report issued by the 9/11 Commission and ensures that money invested in homeland security is done so in a planned and coordinated manner.

National Strategy for Homeland Security

In July 2002, the White House Office of Homeland Security issued the *National Strategy for Homeland Security.* The purpose of the *Strategy* is to mobilize and organize our Nation to secure the U.S. homeland from terrorist attacks. It identifies six Critical Mission Areas, including intelligence and warning, border and transportation security, domestic counterterrorism, protecting critical infrastructure, defending against catastrophic threats, and emergency preparedness and response.

The *Strategy's* objectives are to prevent terrorist attacks within the United States; reduce America's vulnerability to terrorism; minimize the damage and to quickly recover from attacks that do occur.

Establishment of OHS

The California Office of Homeland Security was established in the Office of the Governor by Executive Order. The Office is charged with developing and coordinating a comprehensive state strategy that includes prevention, preparedness, response and recovery. It is responsible for coordinating homeland security activities throughout California. Under this Executive Order, all state agencies and departments are directed to cooperate with the OHS. Some of the key state partners include the Office of Emergency Services, the State Military Department, the California Department of Justice, the California Highway Patrol, Department of Health Services, the Department of Food and Agriculture, and the Emergency Medical Services Authority (EMSA).

Overview of OHS and its Core Functions

Under the Executive Order, the Office of Homeland Security is responsible for developing and implementing the State's strategy to combat terrorism. The fundamental components of the strategy are: preventing acts of terror by enhancing awareness and information sharing; strengthening preparedness by building upon California's robust training and exercise program; deterring attacks by protecting critical assets and reducing their vulnerability; and bolstering the capabilities of first responders with federal homeland security grants. Prevention is our highest priority. We must have an information and warning system that can detect terrorist activity before it manifests itself in an attack so that proper preemptive, preventive, and protective action can be taken. California must work in cooperation with federal agencies to complement their information and warning efforts to ensure that there is an integrated system at the federal, State and local levels.

The first core function of OHS is threat assessment and information analysis and sharing. The OHS promotes information sharing between all levels of law enforcement and, where appropriate, other first responders and the private sector. *The 9/11 Commission Report* identified several instances where such information sharing may have been helpful in preventing the 9/11 attacks. California has led the nation by establishing four regional centers and one statewide center where all agencies can work collaboratively to ensure that relevant information and threat assessments are gathered and shared with all appropriate agencies.

The California Office of Homeland Security, in cooperation with the California Highway Patrol and the California Department of Justice has established the State Terrorism Threat Assessment Center (STTAC) and is using the Law Enforcement and Terrorism Prevention Program funds to establish and operate four Regional Terrorist Threat Assessment Centers, aligned with the four FBI Field offices and Joint Terrorism Task Forces (JTTF) in California. The STTAC coordinates the ongoing information sharing and prevention efforts of State agencies, including the Office of Homeland Security, California Department of Justice, California Highway Patrol, Office of

Emergency Services, Emergency Medical Services Authority, California Department of Food and Agriculture, and other State agencies.

The four Regional Terrorist Threat Assessment Centers provide crucial links and processes to assess emerging threats and effectively disseminate and share timely information. The state system links federal, state and local law enforcement and public safety agencies and ensures that critical information is better shared between law enforcement, fire and emergency services, public health, agriculture commissioners, private security companies and industries. Our state information sharing system will improve the level of critical information provided to our public safety agencies statewide, making them better informed and more successful to deter, detect and prevent terrorism and better prepared to effectively respond to an event if one does occur. The State has also created a network of trained Terrorism Liaison Officers (TLOs) consisting of over 400 public safety officers to better identify the precursors to terrorist attacks.

The second core function is critical infrastructure protection. In December of 2003, the President issued Homeland Security Presidential Directive 7 (HSPD-7) on Critical Infrastructure Identification, Prioritization, and Protection. As HSPD-7 acknowledges; terrorists seek to destroy, incapacitate, or exploit critical infrastructure and key resources across the United States to threaten national security, cause mass casualties, weaken our economy and damage public morale and confidence. OHS recognizes that California is not exempt from this threat and therefore, critical infrastructure protection remains a top priority.

OHS is coordinating with the Federal Department of Homeland Security in conducting site assessments across California's critical infrastructure/key resource sectors within California. OHS coordinated the FY05 Buffer Zone Protection Grant Program (BZPP), which provides funding for equipment and protective measures to better protect, secure, and reduce the vulnerabilities of federally identified critical infrastructure and key resource sites. The critical infrastructure sectors identified by the federal government include: Agriculture and Food; Banking and Finance, Chemical and Hazardous Materials Industry; Defense Industry Base; Energy; Emergency Services; Information Technology; Telecommunications; Postal and Shipping; Public Health; Transportation; Water; and National Monuments and Icons. Key resources include: Commercial Assets; Government Facilities; Dams; Nuclear Power Plants.

The Buffer Zone Protection Program is designed to reduce vulnerabilities of critical infrastructure and key resource sites by extending and hardening the protected area around a site into the surrounding community and supporting the prevention, preparedness, and response efforts of local first responders. Local law enforcement develops and implements buffer zone protection plans to increase the level of protection and act as a deterrent and prevention mechanism for possible terrorist's threats or incidents. The total funding allocation for the FY05 BZPP grants is $91.3 million nationwide. California's allocation is $12.9 million.

A critical component of California's strategy to secure infrastructure is protecting the State's maritime assets. The Office of Homeland Security is a member of the three Area Maritime Security Committees in California. These bodies, chaired by US Coast Guard Captains of the Port, conduct regional threat and vulnerability assessments, develop specific scenarios based upon these assessments, and conduct training and exercises to prevent and respond to a wide variety of incidents. In addition to the $33 million in grants California received under the Port Security Grant Program, the State awarded an additional $5 million to 11 ports from the State's own share of the Homeland Security Grant Program. These grants will be used to prevent terrorists from using improvised explosive devices, as well as investing in training port security personnel, communications equipment, physical security improvements such as cameras, lighting, fencing, underwater surveillance, and personal protective equipment for port first responders.

The third core function is the statewide Homeland Security Training and Exercise Program. This multi-agency, multi-disciplinary team that ensures training and exercises for California's emergency responders are systematically developed and coordinated to respond and recover from terrorist attacks using weapons of mass destruction. The Exercise Program focuses on a robust annual statewide exercise, Golden Guardian, which will take place in less than a month. The Golden Guardian exercise will involve more than 2500 participants representing more than 120 federal state and local agencies. The Golden Guardian 2005 scenario involves simultaneous attacks on critical infrastructure at sites in the San Francisco/Oakland Bay Area and State Capitol regions.

The Golden Guardian Exercise Series was first introduced in California in 2004, and has become an annual exercise conducted to coordinate response mechanisms of city, county and state governmental entities, private sector and volunteer organizations in response to potential man-made events and natural disasters. The goal of the Golden Guardian Exercise Series is to build upon the lessons learned from this and subsequent exercises conducted throughout the nation, as well as real-world events.

The training program also coordinates the States emergency responder training; developing training, ensuring training is properly certified, and tracking those who have been trained. Both programs ensure that the lessons learned from previous exercises are folded back into the next year's exercise objectives and are used to identify new training needs. Additionally, the training and exercise program receives recommendations and guidance on training curriculum development from the Emergency Response and Advisory Training Committee (ERTAC), which was established by legislation in 2003.

Another core function is grant distribution and management. California has been awarded over $1 billion in federal funding since 9/11, in a variety of grant programs. California has designed a grant management system that promotes regional planning and multi-discipline coordination. This is true for grants managed by the Office of Homeland Security, the Department of Health Services and the Office of Emergency Services. The oversight bodies for each of these grants at the State level contains representatives from all the disciplines, including: fire, law enforcement, public health, emergency medical and emergency management. This has ensured that all the grants are receiving input and coordination to promote multi-discipline planning and coordination.

I am confident that California is more prepared today than ever. Working in concert with our federal, State, local and private sector partners we have markedly enhanced capabilities across our core functions of threat assessment and information analysis, critical infrastructure protection, training and exercise, and grant distribution and management. While much has been accomplished, we recognize that improvements can still be made. I appreciate this opportunity to share with the Committee the accomplishments we have achieved and invite suggestions for further development.

Thank you for your attention this morning.

Before the U.S. Senate Judiciary Subcommittee on Terrorism, Technology and Homeland Security
U.S. Senator John Cornyn (R-TX)
"Terrorism: Emergency Preparedness"
Wednesday, October 26, 2005, 10:30 a.m., Dirksen Senate Office Building Room 226

Mr. Chairman, Senator Feinstein, thank you for convening this important hearing. Issues related to emergency preparedness in the event of a terrorist attack must remain a top priority for the Congress. And I agree with you both that this is a timely and important hearing ***particularly*** in light of the natural disasters that affected Texas, Louisiana, Mississippi, and most recently Florida, and our government's response to them. Congress and others throughout the country are in the midst of evaluating the response to Hurricanes Katrina, Rita, and now Wilma, and we should learn from these events in order to better prepare for future emergencies.

But observing the failures of government at all levels following Hurricane Katrina concerns me. I am concerned not only because these failures left many people stranded for a substantial amount of time, but also because it sends a troubling message about our ability to respond to potential terrorist attacks – and this, more than four years since the deadly attacks of September 11th. As we learned in the wake of those attacks, our nation's first responders are the men and women serving on the ground in local police departments, fire departments and emergency rooms. These are the same men and women who play a vital and responsive role during natural disasters, and impediments that prevent them from fully carrying out their duties must be removed.

Unfortunately, many of the deficiencies identified by the 9/11 Commission remain even today. One of the paramount failures analyzed in the Commission's report concerns barriers to effective communication among first responders and other emergency personnel.

This lack of ability to communicate was, unfortunately, again on display when dealing with Hurricane Katrina. Local officials could not communicate with officials in neighboring jurisdictions, and there were impediments to communications between local and federal officials. We must tear down these technical barriers.

Another area identified by the 9/11 Commission that Congress must address is the allocation of our scarce homeland security resources. Specifically, the 9/11 Commission Report recommended that homeland security assistance should be based strictly on an assessment of risks and vulnerabilities. In particular, this assistance should not be a program for general revenue sharing, but should be applied based on risks and vulnerabilities that merit additional support. I agree with this recommendation.

I have worked with Senator Feinstein to address this issue, and we have developed legislation that would require that Federal Homeland Security funds be allocated to states according to a risk-based assessment. It is vital that we better allocate our limited resources to the vulnerable places in the country we most need to protect, and that these funds are distributed in an efficient and timely manner because taxpayer dollars are not limitless.

Congress must diligently work to ensure every penny of homeland security money be directed where it will do the most good -- and that is to places across our nation where terrorists are ***likely*** to strike and where such strikes could do the ***most*** damage to our people, our government, and our national economy.

I recently toured several Texas seaports and visited with port directors, industry leaders, and emergency responders in and around the ports of Houston, Beaumont, and Corpus Christi.

These kinds of facilities and the communities around them have enormous security needs, and the consequences of a terrorist attack on any of these facilities would be devastating, not only to the local communities, but to the economic engine of the whole country. A risk-based homeland security funding mechanism would go far to protecting these types of critical infra-security assets.

And finally, I think it is important that we continue to examine the continuity of our government, and the problems that may arise from a successful mass terrorist attack. As the 9/11 Commission report recognized, a catastrophic attack could occur with little or no notice, and we need to take steps to minimize the possible disruption of national security policymaking. I agree with that assessment but I think we should go farther.

It is believed that United Airlines Flight 93 was likely headed for the U.S. Capitol. But for a late departure and the ensuing heroism of the passengers onboard, our Congress might have suffered catastrophic loss.

Had this happened, there would be a very real concern about whether any of the important legislation that was needed immediately following these attacks would have been enacted in timely fashion.

I have introduced legislation that would provide for a constitutional amendment to quickly reconstitute Congress in the case of a national disaster and to clarify the presidential succession lines, and I hope it will receive serious consideration.

The failure to remedy the deficiencies identified by the 9/11 Commission and others following extensive reviews is simply inexcusable. I look forward to discussing these issues with our witnesses today.

Thank you Mr. Chairman.

REPORT TO CONGRESS

ASSESSMENT:
THE TERRORISM RISK INSURANCE ACT OF 2002

JUNE 30, 2005

THE UNITED STATES DEPARTMENT OF THE TREASURY
OFFICE OF ECONOMIC POLICY
WASHINGTON D.C.

Chapter 1 Introduction and Executive Summary

Introduction

The Terrorism Risk Insurance Act of 2002 (P.L. 107-297) (TRIA) requires the Department of the Treasury to administer a temporary program providing a Federal backstop for specified losses that are covered by insurers in the event of an act of terrorism committed by or on behalf of foreign person or interests. The Act further mandates that, as administrator of the Terrorism Risk Insurance Program (Program), the Treasury Department assess features of the Program and its environment. Specifically, Treasury is required to assess:

- The effectiveness of the Program;
- The likely capacity of the property and casualty insurance industry to offer insurance for terrorism risk after termination of the Program; and,
- The availability and affordability of such insurance for various policyholders, including railroads, trucking, and public transit.

We evaluate the effectiveness of TRIA within the context of the purpose of the legislation. The Act states:

> PURPOSE- The purpose of this title is to establish a temporary Federal program that provides for a transparent system of shared public and private compensation for insured losses resulting from acts of terrorism, in order to--
>
> (1) protect consumers by addressing market disruptions and ensure the continued widespread availability and affordability of property and casualty insurance for terrorism risk; and
>
> (2) allow for a transitional period for the private markets to stabilize, resume pricing of such insurance, and build capacity to absorb any future losses, while preserving State insurance regulation and consumer protections.

Therefore we evaluate effectiveness in terms of these purposes.

We evaluate the likely capacity of the property and casualty insurance industry to offer insurance for terrorism risk after expiration of the Program in terms of the general "insurability" of terrorism risk and the likely provision of coverage after the expiration of TRIA.

We address the "availability and affordability of such insurance for various policyholders, including railroads, trucking, and public transit," by reporting on terrorism insurance take-up and costs for these groups.

To assist in making these assessments, Treasury conducted a set of surveys. Treasury contracted with an outside survey research firm to assist in the development of and to conduct the surveys of both policyholders and insurers. In the process of developing the survey instruments, Treasury worked closely with policyholder and insurance industry representatives.

In addition to the formal surveys, Treasury also consulted with the National Association of Insurance Commissioners (NAIC), and a broad range of experts representing the insurance industry, policyholders, and others, in order to draw upon as many sources of information and input as possible. The completed survey results and information derived from these other sources forms the basis of this Report to Congress.

The Report is organized as follows. Chapter 2 discusses background, including the mechanics of the Program. Chapter 3 gives a methodological overview of the surveys conducted by the Department. Chapter 4 summarizes and discusses key results from the surveys of insurers. Chapter 5 summarizes and discusses key results from the surveys of policyholders. This chapter also discusses the "availability and affordability" of terrorism risk insurance coverage for three specific policyholder groups as mandated by the Act – railroads, trucking, and public transit. Chapter 6 summarizes information collected from the survey of insurers on the use of reinsurance for terrorism risk. In Chapter 7, we make an assessment of the likely capacity of the insurance industry to offer insurance for terrorism risk after expiration of the Program. In making this assessment we use information from the surveys along with industry financial statistics, information on terrorism modeling methodologies, and other types of information, including that derived from our consultations with experts from the insurance and reinsurance industries and insurance industry regulators.

Executive Summary

Based on our research, we offer our assessment of TRIA. Treasury is required to assess:

A. The effectiveness of the Program;
B. The likely capacity of the property and casualty insurance industry to offer insurance for terrorism risk after termination of the Program; and,
C. The availability and affordability of such insurance for various policyholders, including railroads, trucking, and public transit.

A. Effectiveness

As discussed above, we evaluate the effectiveness of TRIA in terms of the purposes given in the legislation.

Item 1: Protect consumers by addressing market disruptions and ensure the continued widespread availability and affordability of property and casualty insurance for terrorism risk.

The year 2003 marked the first of three full years of subsidized Federal terrorism risk reinsurance. Surveys of insurers and policyholders through early 2005 suggest that policyholder take-up and the number of insurers writing terrorism risk insurance improved somewhat between 2002 and the period subsequent to enactment of TRIA (2003 through 2005). TRIA's insurer deductibles have increased each year since 2003, shifting ever more of the burden of coverage from the Federal government back to the industry. In spite of this shift, the data show policyholder take-up rose or stayed stable between 2003 and 2004 and between 2004 and early 2005, a pattern suggestive of a market increasingly able to provide coverage.

Availability of Terrorism Coverage

Insurer Survey Results – Insurers wrote terrorism risk coverage on 67 percent of commercial property and casualty insurance policies, a 7 percentage point increase from 2002. This increase is in part due to more insurers writing coverage for terrorism risk. Whereas 73 percent wrote coverage in 2002, fully 91 percent of insurers surveyed wrote terrorism insurance in 2003. The measures of terrorism risk insurance were generally stable between 2003 and 2004.

Policyholder Survey Results – Between 2002 and 2003, after the enactment of TRIA, take-up of terrorism risk insurance increased from 27 percent of policyholders to 39.5 percent. In 2004, 54 percent of policyholders reported having terrorism risk insurance coverage.

We note that these changes were correlated with the enactment of TRIA, but not necessarily a result of TRIA. Other factors, such as the general insurance underwriting cycle, likely had some effect on market outcomes during this time. While we cannot quantify the share of observed coverage changes caused by TRIA, we do show evidence that suggests some effect of the Federal subsidy.

Pricing of Terrorism Coverage

Results from both the survey of insurers and the survey of policyholders suggest insurers resumed, or more accurately began, pricing terrorism risk insurance during the time TRIA was in effect.

Insurer Survey Results – In 2002, over 75 percent of insurers providing coverage for terrorism risk did not charge for it. That share declined to 46 percent in 2003 and just over 40 percent in 2004. As a consequence of more insurers charging for coverage during this period, the average cost of terrorism insurance (measured as the share of total premiums paid for terrorism coverage) increased from 0.9 percent to 1.8 percent of premiums by 2004.

Among insurers who charged for terrorism insurance in 2002, however, average cost did not follow a straight pattern. The share of premiums charged for terrorism coverage first declined from 3.7 to 2.4 percent of premiums between 2002 and 2003, but then increased to 3.1 percent of premiums by 2004.

Policyholder Survey Results – In 2002, 70 percent of policyholders *with* terrorism risk insurance coverage reported that they received the coverage at no cost. That share declined to 42 percent in 2003 and further to 37 percent in 2004. As a consequence of more policyholders paying for terrorism risk insurance, the average cost of such coverage increased from 1.2 percent of premium in 2002 to 1.6 percent in 2003, and further to 1.7 percent of premium by 2004.

Among policyholders who reported paying for terrorism coverage, cost declined steadily over the period: from 4.0 percent of premium in 2002 to 2.8 percent in 2003 and further to 2.7 percent of premium in 2004.

Policyholders located in high-risk cities faced declining costs for terrorism risk coverage that varied from 2.8 percent of premiums in 2002, 3 percent in 2003 and 1.9 percent in 2004. This overall decline in the cost of terrorism coverage is the outcome of two opposing trends: an increasing share that are paying for coverage and declining prices among those who report paying for the coverage. More than half of policyholders in cities that are considered to be at high risk for a terrorist attack reported receiving coverage at no cost in 2002, but less than 30 percent reported free coverage in 2004. On the other hand, the cost of terrorism coverage for paying policyholders in these high-risk cities declined substantially from 6.1 percent of premium in 2002, to 5.1 percent in 2003, and further to 2.6 percent in 2004.

Item 2: Allow for a transitional period for the private markets to stabilize, resume pricing of such insurance (described above), and build capacity to absorb any future losses, while preserving state insurance regulation and consumer protections.

Building Capacity to Absorb Future Losses

Industry surplus, a major source of insurer capacity, has returned to pre-September 11th levels. Insurers are financially stronger and more able to bear unexpected losses than they were prior to the enactment of TRIA.

Insurers might have begun charging for terrorism risk insurance, and insurer financial strength would have improved whether or not TRIA was enacted. We therefore cannot determine that TRIA effectively caused these changes to take place.

Reinsurance is another important component of an insurer's capacity to absorb losses. Our data show a modest net increase in use of reinsurance over the period. Seventy percent of insurers reported purchasing reinsurance for terrorism risk in 2003, this fell to 65 percent in 2004 before increasing to 75 percent in the first months of 2005. Smaller and medium-sized insurers generally reported greater use of reinsurance for terrorism risk exposure (TRIA deductibles and co-payments) between 2003 and 2005. During this same period, however, larger insurers reported less use of reinsurance for terrorism risk exposure.

Assessment (Items 1 and 2)

Overall we find that TRIA was effective in terms of the purposes it was designed to achieve. TRIA provided a transitional period during which insurers had enhanced financial capacity to write terrorism risk insurance coverage. While we don't ascribe a causal effect, during this period insurers began pricing for terrorism coverage and insurer financial strength improved. More generally, TRIA provided an adjustment period allowing both insurers and policyholders to adjust to the post-September 11th view of terrorism risk.

TRIA's effectiveness for these purposes does not imply continuation of the program. The sunset of TRIA should encourage the development of the private reinsurance market and other risk-transfer mechanisms.

B. Insurance for Terrorism Risk after Termination of the Program

The likely capacity of the property and casualty insurance industry to offer insurance for terrorism risk after expiration of the program is the primary focus of Chapter 7 of the report. TRIA provided a Federal backstop for terrorism losses that effectively subsidized terrorism risk insurance. It is reasonable to expect that the removal of the subsidy will result in a short-lived adjustment in coverage and pricing. We also sketch briefly the likely dynamics of the long run adjustment.

Modeling

To provide and price insurance efficiently, insurers should be able to quantify their exposure to losses from terrorism risk. Insurers' primary tool for quantifying loss exposure is modeling terrorism risk. Modeling terrorism risk has two critical components: (1) the ability to identify and quantify the severity of an event in terms of insurers' losses, and (2) the probability of the loss occurring. Our assessment of developments in risk modeling over the past few years is positive, but we note that challenges do remain.

Insurers' ability to identify and quantify the severity of an event in terms of insurers' losses has improved greatly. In particular, insurers are much better able to assess their exposures or accumulations of risk for a given terrorist event on an overall and individual customer basis. This is important because it allows insurers to more effectively underwrite coverage. Nevertheless, challenges remain, particularly in assessing the probability of the loss from terrorism. Because of the difficulty inherent in assessing these probabilities, use of models to predict terrorism risk is tempered by the uncertainty of their predictions.

Financial Capacity

We also assess capacity in terms of insurer financial strength, which incorporates both balance sheet strength and operating performance. The financial health of insurers, especially surplus, improved in the past three years. Among insurer groups providing coverage in TRIA-eligible property and casualty lines, surplus was higher in the third quarter of 2004 than it was in the third quarters of 2001, 2002 and 2003.

Our surveys of insurers and policyholders also provide some indications of the development of private market capacity. Take-up of terrorism risk insurance, for example, continued to increase, while the ratio of policies written by insurers including terrorism coverage has been flat to rising, even as the TRIA deductible rose over time.

It would not be surprising if the expiration of TRIA changes industry behavior since the business environment will change. Insurers, for example, will likely consider factors such as the possibility of insolvency from terrorism losses given the levels of surplus available and the effect on credit ratings. Experience with natural catastrophe risk underwriting and assignment of agency ratings suggests that in order to avoid ratings downgrades, insurers may significantly alter their approach to terrorism risk insurance after TRIA's expiration. Among the changes insurers may institute are increasing the use of private reinsurance, building surplus by tapping into capital markets, and raising premiums or placing exclusions on some policies.

The policyholder and insurer surveys include direct responses on the availability of coverage after the expiration of TRIA. Responding to questions about policies written in early 2005 that continue into 2006, nearly 50 percent of insurers reported that they are not writing coverage for terrorism risks in 2006 (after the scheduled expiration of TRIA) that is similar to the coverage they write under TRIA. One quarter of policyholders with terrorism risk coverage indicated that their coverage excludes terrorism coverage after the expiration of TRIA.

Assessment

TRIA's expiration will conclude the transitional assistance first provided to the insurance markets in the uncertain economic environment of 2002. Overall, our assessment is that the immediate effect of the removal of the TRIA subsidy is likely to be less terrorism insurance written by insurers, higher prices and lower policyholder take-up. While TRIA is in effect, however, it crowds out development of some reinsurance markets, and delays the development of private capacity to provide terrorism risk insurance. Over time, we expect that the private market will develop additional terrorism insurance capacity. We anticipate that the initial response of premiums in the market will spur the buildup of surplus as insurers tap into capital markets and the development of additional private reinsurance and other risk shifting mechanisms.

C. Availability and Affordability of Such Insurance for Various Policyholders, Including Railroads, Trucking, and Public Transit

Chapter 4 discusses terrorism risk insurance take-up and cost for various classes of policyholders, including railroads, trucking and public transit. Among railroads, estimated take-up rates were approximately 25 percent in 2002, 31 percent in 2003, and 32 percent in 2004. In trucking, we estimate that take-up rates increased from 23 percent in 2002 to 31 percent in 2003 and 43 percent in 2004. The corresponding estimates for public transportation are 2 percent in 2002, 20 percent in 2003, and 36 percent in 2004.

Cost estimates for these populations are less precise because they are based on smaller sample sizes (only respondents with terrorism coverage provide cost information), and are more likely to suffer from selection bias. The cost estimates for railroads are 5 percent of premiums in 2002 and 2003, increasing to 8 percent of premiums in 2004. These estimates, however, are based on only 13-15 observations (out of a population of 421 railroads). Cost estimates for trucking, where sample sizes are reasonably good, fall from 0.68 percent in 2002 to 0.55 percent in 2003 and rise to 1.06 percent in 2004. Cost estimates for public transit are based on samples too small to report.

9/11
Public Discourse Project

Prepared Statement of Senator Slade Gorton

Member, National Commission on Terrorist Attacks Upon the United States

Subcommittee on Terrorism, Technology, and Homeland Security
Committee on the Judiciary
United States Senate
October 26, 2005

Chairman Kyl, Ranking Member Feinstein, Members of the Subcommittee:

- In the wake of Katrina, it is clear that in the area of emergency response we have not heeded the lessons of 9/11. Today I will discuss briefly what those lessons are, and what we must do to avoid having to learn them a third time through yet another inadequate response.

- I thank the Committee for inviting me here today.

- I specifically commend Senator Kyl and Senator Feinstein for holding this hearing and performing this oversight. Your attention, and that of other Committees of the Congress, will be a key to getting the serious problems in this area resolved.

- As we learned on 9/11, the threats the American people face are not confined to distant battlefields—they can materialize here at home.

  If terrorists strike again on American soil, it will be local emergency responders—police, firefighters, and emergency medical technicians—who will be on the front lines. Local emergency preparedness is now a matter of national security. In addition, of course, while the federal government through FEMA is not generally a first responder, its utterly inadequate response to the needs of both victims and first responders to Katrina calls for dramatic changes in its preparation for, and response to, both natural and terrorist-caused emergencies.

COMMUNICATIONS FAILURES

- On 9/11 shortcomings in emergency communications hindered first responders and led to unnecessary loss of lives. The problem was especially bad:

- o Among firefighters, in the Twin Towers; and
- o Between agencies responding to the World Trade Center site.

- As those heroic firefighters in both towers climbed higher, their radio transmissions were disrupted by the many floors between them and their commanders. Communications with their chiefs in the lobby became weaker and more sporadic.
  - o Because so many people were trying to speak at once, available channels were overwhelmed. Transmissions overlapped and often became indecipherable.
  - o Many firefighters in the North Tower didn't hear the evacuation order issued after the South Tower collapsed. Some weren't even aware that the South Tower had come down.
- Meanwhile, communications among agencies were extremely poor.
  - o In one well-known example, fire chiefs in the lobbies of the towers got no information from the police helicopters circling above.
- Because of poor interagency communications, many redundant searches were conducted that morning. This wasted precious time and caused the deaths of many heroic first responders.
- Hurricane Katrina reminds us that this problem has not been solved. In Katrina, poor public safety communications again delayed the response.
  - o New Orleans and three neighboring parishes were using different equipment and different frequencies—they couldn't talk to one another.
  - o Helicopter crews couldn't talk to rescuers in boats.
  - o National Guard commanders in Mississippi had to use human couriers to carry messages.
- Last July the 9/11 Commission recommended that Congress turn over broadcast spectrum to first responders, to improve communication within

agencies and allow interoperability among agencies.

- The House and Senate are finally moving forward on legislation to reclaim analog TV spectrum, currently held by broadcasters, and to designate some of it for use by emergency responders.

    - But the date in the bill just released by the Commerce Committee is April 7, 2009—nearly 8 years after the 9/11 attacks. This is far too long.

    - By contrast, less than four years after Pearl Harbor both Japan and Germany had been defeated. It is ridiculous that it should take eight years to implement such an obvious response to the 9/11 attacks.

    - Experts say that this transition could be accomplished as early as 18 months from today and certainly within two years.

    - There will surely be another terrorist attack or major disaster in the next four years. We need a sense of urgency to get this done <u>now</u>—not in four more years.

COMMAND AND CONTROL

- On 9/11 in New York and in New Orleans, command structures for emergency response were not clearly defined. It was not clear beforehand who was in charge, or what each agency's responsibilities were. This confusion cost lives.

- I have the impression that Mississippi's response to Katrina did not suffer from the same problems of command and control as did that of Louisiana. Command and control in response to Hurricane Rita seems to have worked somewhat better as well.

    - The Committee may wish to examine the facts and circumstances of command and control in these cases, so that we can learn from them.

- The 9/11 Commission recommended that local governments adopt the Incident Command System. This system defines who is in charge and what agencies' responsibilities are in a crisis.

- o Every locality should have a clear emergency plan, with every agency's specific role laid out beforehand, in black and white.

- o As we saw in Katrina, if local plans are not highly specific and are not regularly rehearsed, confusion is inevitable.

- o DHS set a hard deadline of October 1, 2006 for localities to establish and exercise a command and control system to qualify for first-responder grants.

    - That deadline must not slip.

    - Localities that do not have clear, well-rehearsed incident command plans by that date should not receive federal homeland security grants.

RISK-BASED FUNDING

- Since 2001, the federal government has allocated more than $8 billion to help state and local governments prepare for terrorist attacks. Unfortunately, these funds have not been guided by any assessment of risk and vulnerability.

- To solve this problem, the Commission made a common-sense recommendation: that federal homeland security assistance be based strictly on an assessment of risks and vulnerabilities, not politics.

    - o These funds are national security funds. They should not be subject to revenue sharing.

    - o These funds are too important to be spent without any guarantee that they are actually reducing our vulnerabilities, or increasing our overall security from terrorism.

    - o The Administration supports this reform; Chairman Kyl and Ranking Member Feinstein support this reform, as do many other Senators.

    - o The House has attached an excellent bipartisan risk-based formula proposal to the PATRIOT Act reauthorization bill. We urge the Congress to adopt that proposal and solve this problem this year.

RISK ASSESSMENT

- The Intelligence Reform Act required DHS to produce a National Strategy for Transportation Security by April 1, 2005.

    - The Senate finally received that report last month.

    - Unfortunately, the strategy remains classified. As such it is unavailable to the public, the transportation community, state and local governments, and first responders.

    - This report will be of little practical use until it can be distributed to those responsible for its implementation.

- DHS has still not produced the national risks and vulnerabilities assessment for critical infrastructure, which was due June 15th.

    - Until this report is completed it will be impossible to allocate homeland security funds in a rational manner, based on risks and vulnerabilities.

- Finally, this type of assessment needs to be an ongoing process, not a one-time job. DHS should be able to modify this calculus as the threat environment and our state of readiness change.

CONCLUSION

- As Hurricane Katrina reminded us, large-scale emergency responses are bound to occur again in the future, whether from terrorist attacks or natural disasters.

- Mr. Chairman, the question is: Are we better prepared for the next major terrorist attack? For the next natural disaster? Are we prepared for an attack with a dirty bomb, or one with chemical or biological weapons?

    - Are our emergency communications good enough?
    - Are our response plans updated and rehearsed?
    - Are we directing federal funds where they are needed?

        - to protect our greatest vulnerabilities?

        - to meet the most catastrophic threats?

- Have we systematically assessed what our greatest vulnerabilities are?

- Today, sadly, the answer to all of these questions is no. After 9/11, after Katrina, we are still not prepared.

- We look forward to working with you, and with your counterparts in the House, to enact these common-sense recommendations into law this year—for the safety of our first responders, and the communities they are pledged to protect.

- The lessons of 9/11, and again of Katrina, are too painful to be learned a third time.

- I thank the Chairman and Ranking Member for your continuing attention to these, and other, important national security issues. I would be pleased to respond to your questions.

###

Statement
United States Senate Committee on the Judiciary
Terrorism: Emergency Preparedness
October 26, 2005

The Honorable Patrick Leahy
United States Senator , Vermont

Statement of Senator Patrick Leahy
Hearing Before the Subcommittee on
Terrorism, Technology & Homeland Security
"Terrorism: Emergency Preparedness"
October 26, 2005

I thank Senator Kyl for convening today's hearing and commend his efforts to address the critical issue of terrorism preparedness. I want to welcome each of our witnesses, particularly my friend Slade Gorton.

Terrorism was not a top priority of the Bush Administration when it took office in January 2001. Problems ranged from an understaffed foreign translation program, woefully inadequate information systems, and cultural attitudes that frustrated information-sharing across agencies. Just one day before the September 11, 2001, terrorist attacks, the serving Attorney General rejected the FBI's request to include more money for counterterrorism in his budget proposal. After the attacks of 9/11, the Administration resisted this Committee's efforts to examine what led to the tragedy. The Administration fought attempts by Congress to establish a Department of Homeland Security. It tried to block the formation of the 9/11 Commission, and then it put roadblocks in the way of the Commissioners as they worked to fully investigate under its mandate.

Unfortunately, this Administration has displayed a clear pattern by misreading the warning signs of impending disaster, whether in the form of a terrorist attack or a natural disaster, often with dire consequences for Americans. We recently saw the Administration's failure to plan for, and quickly react to, Hurricane Katrina. It has been shocking and sobering to see how woefully unprepared our government was to help the victims of this disaster. The government failed to react despite reports that were prepared on exactly the point at hand – a hurricane in New Orleans that caused the levees to breach. The chaotic response raises serious concerns about the adequacy of our preparedness to respond to terrorist attacks, despite the many billions of dollars spent by the Department of Homeland Security since 9/11.

Terrorism experts warn about the possibility of a catastrophic bioterrorist attack, yet we learned last fall that we are not prepared to meet the biological threat that comes every year – influenza. I hoped that the Bush Administration would learn a lesson from the 2003 shortage of flu vaccines. Instead, health providers across the country, including in my home state of Vermont, were forced to ask healthy people to forgo their flu shot. If the government's top health officials can not prepare for the seasonal flu – an annual occurrence – what does that portend about their ability to prepare for biological terrorist attacks?

The President only recently addressed the threat of avian flu spreading to the United States, but the Administration's track record gives me little confidence that an outbreak in our country would be handled effectively. I am particularly troubled by the President's proposal to use the military to enforce quarantines and travel restrictions in the event of an avian flu outbreak. Putting the military in the lead role in domestic disaster situations negates fundamental attributes of the National Guard, its experience in working with local and state first responders and its familiarity with local communities.

It would be comforting if we could at least tell Americans that their Government was doing everything possible and practical to keep them safe. Unfortunately, we cannot truthfully tell them that. As we sit here today, there is much left undone in securing our nation. The 9/11 Public Discourse Project, a public education campaign founded by the members of the 9/11 Commission, issued a report in September on our government's progress towards improving emergency preparedness and response. The report evaluates five recommendations issued by the 9/11 Commission in July 2004. The progress of each recommendation received a status ranking; four received a "minimal progress" ranking and the fifth received a ranking of "unsatisfactory." Commissioners Kean and Hamilton issued a statement with the report in which they warned that "Congress and the Executive branch need to step up to the plate. They need to respond with a necessary sense of urgency to adopt the reforms we recommend on emergency preparedness and response."

While I share the concerns of the 9/11 Commissioners, I am hesitant to fully embrace one of their recommendations on emergency preparedness. The Commission recommends basing federal homeland security funding strictly on an assessment of risks and vulnerabilities. I strongly believe that every state – rural or urban, small or large – has basic domestic security needs and deserves to receive federal funds to meet those needs. Instead of pitting large states against small states, as the Administration has done by shortchanging overall resources for first responders, the needs of both should be recognized and addressed. These funds help police, fire and rescue squads meet the homeland security responsibilities the federal government is asking them to meet.

The attacks on 9/11 were a horrible tragedy for our nation. They should have also served as a wake up call. Four years later we remain unprepared for another major terrorist attack. Our inadequate response to natural disasters highlights how vulnerable we are to a major act of terrorism. Rather than hope that an attack does not occur, we should act as if an attack is inevitable and strive to be fully prepared. Our main focus must remain on preventing terrorism, but we cannot dismiss the possibility that we will someday confront another day like 9/11.

Testimony before the Senate Subcommittee on Terrorism, Technology, and Homeland Security of the Committee of the Judiciary, for October 26, 2005

ASSESSING THE STATE OF HOMELAND SECURITY

Michael O'Hanlon, mohanlon@brookings.edu (10/21/05)

Chairman Kyl, Senator Feinstein, and other members of the subcommittee, it is an honor to appear before you today to discuss the enormously important question of our nation's security against possible future terrorist attacks. The recent Katrina experience reminds us of what is at stake, especially since a terrorist attack would clearly provide substantially less warning. President Bush's October 6, 2005 speech in Washington revealed how many times potential terrorist strikes had been in the works against the United States even since 9/11/2001.[1] The July subway attacks in London remind us that the danger of such attacks has not ended, even within the western world. And globally, the strength of the jihadist terror movement (broadly defined) is on balance as great as ever. Clearly we cannot let down our guard. Yet we must also be judicious, cost-effective, and pragmatic in how we attempt to counter terrorism here at home, given the costs to our pocketbooks and way of life of any excessive efforts to protect the homeland.

I have been asked to explore the likely consequences of several potentially severe terrorist scenarios and to assess what steps may have been taken already to address the risks. I will do this with a review of steps taken—and not taken—since 2001, as well as a compendium of tables and quantitative estimates showing the possible human and economic costs of various types of successful attacks. But first, I will summarize several main conclusions.

- The United States has taken a number of impressive steps since 9/11/2001 to protect itself against terror. The greatest progress has been witnessed in air

[1] Reportedly, three attacks were intended for targets on U.S. soil. See Peter Baker and Susan B. Glasser, "Bush Says 10 Plots by Al Qaeda Were Foiled," *Washington Post*, October 7, 2005, p. 1.

security, protection of key government property and prominent infrastructure and other symbolically significant sites in our country, some types of protection against biological attack, elimination of legal and bureaucratic barriers due to the Patriot Act and intelligence reform, and greater integration of our border security agencies as well as our terrorism watch lists.

- However, even within these relatively successful areas, much remains to be done. Private planes are not regulated as well as commercial ones. Large private skyscrapers are not all prudently protected against truck bombs or biological or chemical attacks. Capacity to produce and distribute antidotes to most types of biological attacks is woefully insufficient. Border security resources remain too limited, and intelligence integration cannot yet begin to truly "connect dots" about looming terrorist strikes through automated information analysis.
- And many types of protective measures remain to be even seriously initiated. For example, the chemical industry and the transportation systems that serve it are barely protected at all. Passenger trains and buses are still very vulnerable (perhaps, to some extent, inevitably so). Electricity infrastructure is badly protected and systemically fragile. Food supplies are largely undefended.
- Some types of possible homeland security measures are currently either impractical or unnecessary (or some combination thereof). These include national cruise missile defense, 100% screening of cargo containers entering the country, protection of most malls and restaurants against suicide bombers and individuals with semiautomatic weapons, and creation of large additional hospital capacity for quarantining patients with contagious diseases. But many other measures are overdue, in that they would respond to major national vulnerabilities and do so with good effectiveness at reasonable cost.
- A number of plausible terrorist scenarios could be every bit as bad, if generally not as geographically extensive in effect, as Hurricane Katrina. We should use the reminder of that terrible natural catastrophe to focus ourselves not only on rebuilding the Gulf Coast and improving disaster response, but continuing to improve homeland security more generally with a sense of urgency.

- A specific scenario akin in some ways to the Katrina experience is worth noting. An attack against the Hoover or Glen Canyon dams on the Colorado River could be catastrophic in at least three ways—the rapid inundation of small nearby towns, with high fatality rates likely; the probable destruction of large swaths of major downriver cities, notably Las Vegas; and the extended economic disruption resulting from demolition of facilities so critical to the water and electricity supplies of the southwestern United States.[2]

MAIN TESTIMONY

Since the attacks of September 11, 2001, a good deal has been done to improve the safety of Americans. Much of that increase in safety has come from offensive operations abroad—the military overthrow of the Taliban and associated attacks against al Qaeda, as well as the intelligence and covert operations conducted by the United States in conjunction with key allies such as Pakistan. These steps have lessened U.S. vulnerability to the kind of attacks the country so tragically suffered four years ago.

Homeland security efforts have improved too. Now aware of the harm terrorists can inflict, Americans are on alert, providing a first, crucial line of defense. Air travel is much safer, with screening of all passenger luggage, hardened cockpit doors on all large American commercial aircraft, thousands of air marshals, and armed pilots on some commercial and cargo flights.

Intelligence sharing has improved, especially information about specific individuals suspected of ties to terrorism, through increased integration of databases and greater collaboration between the FBI and the intelligence community. These initial efforts have now been reinforced by the passage of the 2004 Intelligence Reform and Terrorism Prevention Act of 2004. Such linkages of databases can enable offensive operations

---

[2] See for example, "Terrorist Threats Pose Little Inconvenience," January 2004, www.thinkandask.com/news/hooverdam.html.

abroad; they can also assist greatly in the more defensive, but equally critical, domain of homeland security operations.

The share of FBI resources devoted to counterterrorism has doubled, and combined CIA/FBI personnel working on terrorist financing alone have increased from less than a dozen to more than 300 since September, 2001.[3] International cooperation in sharing information on suspected terrorists has improved—extending beyond countries that have been helpful over many years such as France and Britain to include many other states such as Pakistan and Saudi Arabia that now take the threat more seriously.

Additional efforts have also been initiated. A number took place after the 2001 anthrax attacks, others were responses to information gained in prisoner interrogations and other intelligence efforts. Suspicious ships entering U.S. waters are now screened more frequently. The country's exposure to biological attacks has been lessened by stockpiling of hundreds of millions of doses of antibiotics and smallpox vaccine.[4] Oversight rules have been tightened on labs working with biological materials (though actual implementation of those rules, including completion of background checks on lab employees, has lagged).[5] Terrorism insurance is now backstopped by a new federal program. Certain types of major infrastructure, such as well-known bridges and tunnels, are protected by police and National Guard forces during terrorism alerts. Nuclear reactors have better protection than before.[6] Federal agencies are required to have security programs for their information technology networks, and many private firms

---

[3] Vicky O'Hara, "Terrorist Funding," National Public Radio, Morning Edition, November 20, 2003; Speech of George W. Bush at the FBI Academy, Quantico, VA, September 10, 2003; and Philip Shenon, "U.S. Reaches Deal to Limit Transfers of Portable Missiles," *New York Times*, October 21, 2003, p. A1.
[4] Tom Ridge, "Since That Day," *Washington Post*, September 11, 2003, p. 23.
[5] Martin Enserink, "Facing a Security Deadline, Labs Get a 'Provisional' Pass," *Science*, November 7, 2003, p. 962.
[6] There may be some gaps in these types of protective measures to date, but the overall level of security is generally good. See Statement of Jim Wells, General Accounting Office, "Nuclear Regulatory Commission: Preliminary Observations on Efforts to Improve Security at Nuclear Power Plants," GAO-04-1064T, September 14, 2004.

have backed up their headquarters and their databanks so that operations could survive the catastrophic loss of a main site.[7]

We have prepared fairly well to fight the last war--that is, to stop the kinds of attacks that the United States has already experienced. We have done much less, however, to thwart other kinds of plausible strikes. It made sense to move quickly to prevent al Qaeda, with its longstanding interest in airplanes, from easily repeating the 9/11 attacks. But it is time to do a more comprehensive and forward-looking job of protecting the American people.

Al Qaeda may not be as capable as before of "spectacular" attacks in coming years. But it is certainly still capable of using explosives and small arms, with considerable lethality. It may be able to use surface-to-air missiles and other methods of attack as well.[8] There have not been more attacks within the United States. But according to an October, 2005 speech by President Bush, the United States has disrupted three attempted al Qaeda strikes inside the United States, and intercepted at least five more efforts to case targets or infiltrate terrorists into this country.[9] Moreover, the years 2002, 2003, and 2004 have been among the most lethal in the history of global terrorism, with attacks afflicting a wide swath of countries from Spain and Morocco to Tunisia, Saudi Arabia, Pakistan, Indonesia--and of course Iraq.[10] The pattern continued in 2005, and the July 7 London attacks reminded Americans of their continued vulnerability as well.[11]

---

[7] John Moteff, "Computer Security: A Summary of Selected Federal Laws, Executive Orders, and Presidential Directives," Congressional Research Service Report for Congress RL32357, April 16, 2004, p. 2.

[8] David Johnston and Andrew C. Revkin, "Officials Say Their Focus Is on Car and Truck Bombs," *New York Times*, August 2, 2004, p. A13.

[9] President George W. Bush, Speech on Terrorism at the National Endowment for Democracy, October 6, 2005, available at www.whitehouse.gov [accessed October 6, 2005].

[10] See Gilmore Commission (Advisory Panel to Assess Domestic Response Capabilities for Terrorism Involving Weapons of Mass Destruction), Fifth Annual Report, *Forging America's New Normalcy: Securing Our Homeland, Preserving Our Liberty* (Arlington, Va.: RAND Corporation, December 15, 2003), p. 1; Alan B. Krueger and David D. Laitin, "'Misunderestimating' Terrorism," *Foreign Affairs*, vol. 83, no. 5 (September/October 2004), p. 9; and Susan B. Glasser, "U.S. Figures Show Sharp Global Rise in Terrorism," *Washington Post*, April 27, 2005, p. 1.

[11] Richard Benedetto, "Americans Expect Attacks, Poll Finds," *USA Today*, July 12, 2005, p. 1.

A U.N. study in early 2005 argued that al Qaeda continues to have easy access to financial resources and bombmaking materials.[12] There were serious worries that al Qaeda would use truck bombs to destroy key financial institutions in New York, Newark, and Washington in 2004.[13] The "shoe bomber," Richard Reid, attempted to destroy an airplane headed to the United States in 2002.[14] U.S. intelligence reports in early 2005 suggested the possibility of attacks using private aircraft or helicopters.[15] Al Qaeda prisoner interviewers and confiscated documents suggest other possible attacks ranging from blowing up gas stations to poisoning water supplies to using crop dusters to spread biological weapons to detonating radioactive dirty bombs.[16] And the country's chemical industry as well as much of its ground transportation infrastructure remain quite vulnerable, as argued by former Deputy Homeland Security Advisor Richard Falkenrath.[17]

Although al Qaeda has been weakened at the top, it remains extremely dangerous.[18] It is now less of a vertical organization than a collection of loosely affiliated local groups that share motivation--and that, like terrorist groups in general, watch and learn from each other.[19] Former CIA Director George Tenet put it succinctly in 2004: "Successive blows to al-Qaeda's central leadership have transformed the organization into a loose collection

---

[12] Leyla Linton, "Al-Qaeda, Taliban Can Still Launch Attacks, Report Says," *Philadelphia Inquirer*, February 16, 2005.

[13] Eric Lichtblau, “Finance Centers Are Said to Be the Targets,” *New York Times*, August 2, 2004, p. 1.

[14] Shaun Waterman, "Al Qaeda Warns of Threat to Water Supply," *Washington Times*, May 29, 2003, p. 6; and Eric Lichtblau, "U.S. Cites al Qaeda in Plan to Destroy Brooklyn Bridge," *New York Times*, June 20, 2003, p. 1.

[15] Eric Lichtblau, "Government Report on U.S. Aviation Warns of Security Holes," *New York Times*, March 14, 2005, p. A1.

[16] Matthew Brzezinski, *Fortress America* (New York: Bantam Books, 2004), pp. 16-17.

[17] Statement of Richard Falkenrath before the Senate Committee on Homeland Security and Governmental Affairs, January 26, 2005, pp. 12-14.

[18] See Marc Sageman, *Understanding Terror Networks* (Philadelphia: University of Pennsylvania Press, 2004).

[19] The Advisory Panel to Assess Domestic Response Capabilities for Terrorism Involving Weapons of Mass Destruction (Gilmore Commission), *Implementing the National Strategy* (December 2002), p. 11; and Douglas Farah and Peter Finn, "Terrorism, Inc.," *Washington Post*, November 21, 2003, p. 33. On the assertion that modern terrorist groups watch and learn from each other, see Bruce Hoffman, "Terrorism Trends and Prospects," in Ian O. Lesser, Bruce Hoffman, John Arquilla, David Ronfeldt, and Michele Zanini, *Countering the New Terrorism* (Santa Monica, Calif.: RAND, 1999), pp. 8-28; and on the nature of al Qaeda and affiliated as well as sympathetic organizations, see Paul R. Pillar, *Terrorism and U.S. Foreign Policy* (Washington, D.C.: Brookings, 2001), pp. 54-55.

of regional networks that operate more autonomously."[20] There are benefits from dispersing al Qaeda in this way; the near-term risk of sophisticated catastrophic attacks has probably declined as a result. But the risk of smaller and sometimes quite deadly strikes clearly has not--and the possibility of further catastrophic attacks may well increase again in the future.

The benefits gained by depriving al Qaeda of its previous sanctuary in Afghanistan may not be permanent. That organization may ultimately learn to reconstitute itself with a less formal and more virtual and horizontal network. It may also learn how to avoid terrorist watch lists with some effectiveness, for example by using new recruits—including possibly non-Arabs—to conduct future attacks against western countries.[21] The United States is fortunate not to have, as best we can determine, many al Qaeda cells presently on its soil, as several European countries do. It will be challenging, however, to keep things that way.[22]

As the then-Secretary of Homeland Security, Tom Ridge, said in response to a question about whether he was surprised that there hadn't been another attack on U.S. soil since 9/11, "I'm grateful. That's a better way to put it...many things have been done that have altered their environment...But maybe they just weren't ready. They are strategic thinkers. Even if we've altered their environment and our environment here, they aren't going to go away. They're just going to think of another way to go at the same target or look for another target."[23] CIA Director Porter Goss told Congress in February 2005 that "It may be only a matter of time before al Qaeda or another group attempts to use chemical, biological, radiological, and nuclear weapons."[24] DHS has conducted "red cell" exercises involving a diverse range of creative outside thinkers to contemplate

---

[20] Cited in Daniel L. Byman, "Homeland Security: We're Safer Than You Think," *Slate*, August 2, 2004.
[21] "Washington in Brief," *Washington Post*, July 17, 2004, p. A5.
[22] Byman, "Homeland Security," *Slate*, August 2, 2004; and ABC News, "No 'True' Al Qaeda Sleeper Agents Have Been Found in U.S.," abcnews.com, March 9, 2005.
[23] "Ridge 'Grateful' U.S. Has Not Been Hit Again," *USA Today*, August 11, 2004, p. 11; and John Mintz and Sara Kehaulani Goo, "U.S. Officials Warn of New Tactics by Al Qaeda," *Washington Post*, September 5, 2003, p. 2.
[24] Bill Gertz, "Goss Fears WMD Attack in U.S. 'A Matter of Time,'" *Washington Times*, February 17, 2005, p. 3.

possible new ways al Qaeda might attack, but policy responses to such possibilities have typically been limited in scope and scale.[25]

The Iraq war, whatever its other benefits, also appears not to have alleviated the global terrorism problem. In fact, it is quite possible that it has made it worse by aiding al Qaeda's recruiting efforts and providing an opportunity for a core of hardened terrorists to hone their skills and tighten their organizational networks. To quote Goss again, "Islamic extremists are exploiting the Iraqi conflict to recruit new anti-U.S. jihadists. These jihadists who survive will leave Iraq experienced and focused on acts of urban terrorism."[26] The National Intelligence Council reached a similar conclusion in its 2004 report, *Mapping the Global Future*.[27]

It is simply not possible to defend a large, open, advanced society from all possible types of terrorism. The United States contains more than half a million bridges, nearly 500 skyscrapers, nearly 200,000 miles of natural gas pipelines, more than 2,800 power plants--the list of critical infrastructure alone is far too long to protect everything, to say nothing of restaurants and movie theaters and schools and malls.[28] Certain special measures, such as providing tight security and even electronic jamming (against the possibility of GPS-guided munitions attack) around the nation's 104 nuclear power plants, clearly cannot be extended to all possible targets.[29]

But to say that we cannot do everything is not to argue for inaction. There is a strong case for taking additional steps to reduce the risks of catastrophic attacks. Al Qaeda seems to prefer such attacks for their symbolic effects and potential political

---

[25] John Mintz, "Homeland Security Employs Imagination; Outsiders Help Devise Possible Terrorism Plots," *Washington Post*, June 18, 2004, p. A27.
[26] Dana Priest and Josh White, "War Helps Recruit Terrorists, Hill Told," *Washington Post*, February 17, 2005, p. 1.
[27] National Intelligence Council, *Mapping the Global Future* (December 2004), p. 94.
[28] Richard K. Betts, "The Soft Underbelly of American Primacy: Tactical Advantages of Terror," *Political Science Quarterly*, vol. 117, no. 1 (Spring 2002), p. 30.
[29] On jamming, see "U.S. Homeland Defense Strategists," *Aviation Week and Space Technology*, September 6, 2004, p. 20.

consequences; it is also such tragedies that most jeopardize the country's overall well-being.

Catastrophic attacks include, of course, those that cause large numbers of direct casualties. They also include strikes causing few casualties but serious ripple effects, especially in the economic domain. If a nuclear weapon were discovered in a shipping container, for example, casualties might be prevented--but a shutdown in the nation's trade for a substantial period of time could result as policymakers sought means to prevent a recurrence. Or if a shoulder-launched surface-to-air missile took down an airplane, casualties might be modest--depending on the plane, only a few dozen might be killed--but the effects on the nation's air travel could be devastating and longer-lasting than those of September 11, 2001. As another example, the use of a radiological weapon (in which a conventional explosive disperses radioactive material) would be unlikely to kill many, but could require a very costly and time-consuming cleanup.[30]

Even in areas where homeland security has improved, deficiencies often remain. For example, while antibiotic stocks for addressing any anthrax attack are now fairly robust, means of quickly delivering the antibiotics appear still to be lacking.[31] In the domain of air travel, passengers are not generally screened for explosives, cargo carried on commercial jets is generally not inspected, and private airliners face minimal security scrutiny. Perhaps most of all, whatever security improvements have been made for U.S. carriers, fewer have been made to many foreign carriers that transport large numbers of Americans to and from the United States. Moreover, longer-term worries about biological attacks remain acute, since there could be many types of infectious agents for which antidotes prove unavailable when they are most needed. And the private sector has, for the most part, done very little to protect itself.[32]

---

[30] Peter D. Zimmerman with Cheryl Loeb, "Dirty Bombs: The Threat Revisited," *Defense Horizons*, no. 38 (January 2004).
[31] Lawrence M. Wein and Edward H. Kaplan, "Unready for Anthrax," *Washington Post*, July 28, 2003, p. A21.
[32] Statement of Richard Falkenrath before the Senate Committee on Homeland Security and Governmental Affairs, January 26, 2005, pp. 14-15.

It would be a mistake to assume that the creation of the Department of Homeland Security will automatically lead to better protection against such threats. Such reorganizations are extremely difficult, time consuming, and distracting. They can distract attention from efforts to identify remaining key American vulnerabilities and then mitigate them.[33] These problems were of course witnessed during and after Hurricane Katrina in 2005, when FEMA's response to the disaster hardly seemed to have been facilitated by its incorporation within a larger, new organization.

Carrying out a major governmental overhaul when the threat to the nation is so acute is a risky proposition—and not the way the country has typically responded to national crises before. The Department of Defense was not created during World War II, when military leaders had more immediate tasks at hand, but afterwards. Even its much more modest Goldwater Nichols reorganization in 1986 was carried out during a time of relative international peace. By contrast, the DHS has been created in what amounts to a wartime environment—just when its constituent agencies needed to focus on their actual jobs rather than bureaucratic reorganization. Now that that decision has been made, and the third largest department in the government created, it is imperative not to confuse its existence with a successful strategy for protecting the homeland.

And while Congress has improved its ability to address homeland security issues by creating dedicated authorization committees and appropriations subcommittees in both houses, it has not gone far enough. These committees and subcommittees must share jurisdiction with many other committees and subcommittees that insist on a share of the decisionmaking power.[34] This approach is extraordinarily inefficient for executive branch officials who must work with the Congress; in addition, it breeds parochialism among the individual committees and subcommittees about the particular dimensions of homeland security they address.[35] Congress needs to establish the principle that

---

[33] Statement of Richard Falkenrath before the Senate Committee on Homeland Security and Governmental Affairs, January 26, 2005, pp. 2, 7.

[34] For a similar critique of Congress's role, see 9/11 Commission, *The 9/11 Commission Report* (New York: W.W. Norton and Co., 2004), pp. 420-422.

[35] See Statement of Richard Falkenrath before the Senate Committee on Homeland Security and Governmental Affairs, January 26, 2005, p. 4.

homeland security committees and dedicated appropriations subcommittees should have exclusive jurisdiction over funding that is found within the homeland security realm. Cross-jurisdictional input--that is, the need to gain approval of any initiative from more than one authorizing or appropriating body per house of Congress--may in rare instances be appropriate, but should not be the norm.

TABLES AND QUANTITATIVE ESTIMATES

TABLE 1: POSSIBLE SCALE OF TERRORIST ATTACKS

Type of Attack	Possible Fatalities	Estimated Likelihood
Efficient high-potency biological attack	1,000,000	extremely low
Atomic bomb detonated in US city	100,000	very low
Attack (e.g., with conventional explosive or airplane) on nuclear or toxic chemical plant	10,000	very low
Relatively inefficient biological or chemical attack in a stadium, train station, skyscraper	1,000	low
Conventional ordnance attack on train, plane	300	modest
Suicide attack with explosives or firearms in a mall or crowded street	100	modest

Sources: Office of Technology Assessment, *Proliferation of Weapons of Mass Destruction: Assessing the Risks* (Washington, D.C.: U.S. Congress, 1993); and Michael E. O'Hanlon, Peter R. Orszag, Ivo H. Daalder, I.M. Destler, David L. Gunter, James M. Lindsay, Robert E. Litan, and James B. Steinberg, *Protecting the American Homeland: One Year On* (Brookings, 2003), p. 6.

TABLE 2: ECONOMIC DISRUPTION AS A RESULT OF TERRORISM

Type of Attack	Nature of Economic Disruption	Potential Costs
Weapons of mass destruction shipped via container, mail	Extended shutdown in trade, loss of life, physical destruction, lost production in affected area	Up to $1 trillion
Efficient biological attack	Disruption to economic activity in affected area, loss of life, loss of confidence throughout economy	$750 billion
Widespread terror against key elements of public economy across nation (malls, restaurants, etc.)	Significant and sustained decline in economic activity in public spaces, loss of confidence	$250 billion
Attack on interstate natural gas pipelines in southeast US	Natural gas shortages in north and Midwest, significant decline in economic activity in north	$150 billion
Cyberattacks on computer systems regulating electric power combined with physical attacks on transmission/ distribution network	Regional electricity shortages for a week, health risks from heat and cold, interruption of production schedules, destruction of physical capital	$25 billion
Bombings/bomb scares	Effective shutdown of several major cities for a day or two	$10 billion

Source: O'Hanlon et. al., *Protecting the American Homeland*, p. 7

For more tables, please see attached files.

AREAS OF CONTAMINATION FOR RADIOLOGICAL DISPERSAL DEVICES

CESIUM BOMB IN WASHINGTON, DC (LONG TERM CONTAMINATION)

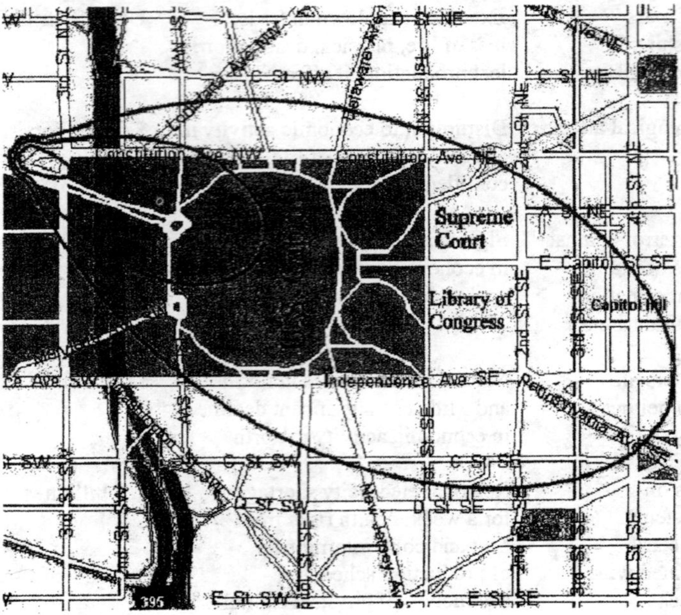

RDD: medical gauge of cesium, dispersed by ten pounds of TNT.
Inner Ring: One cancer death per 100 people due to remaining radiation (5% increase).
Middle Ring: One cancer death per 1,000 people due to remaining radiation (.5% increase): approximately five city blocks.
Outer Ring: One cancer death per 10,000 people due to remaining radiation (.05% increase): approximately one mile swath or 40 city blocks.

COBALT BOMB IN NEW YORK CITY (LONG TERM CONTAMINATION: EPA STANDARDS)

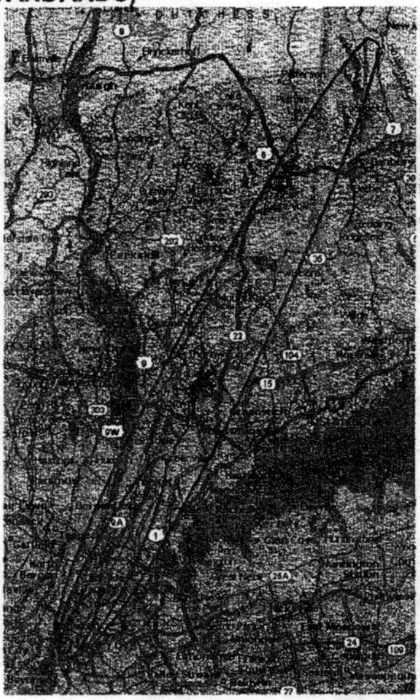

RDD: Single piece of radioactive cobalt from food irradiation plant (approximately one inch by one foot) dispersed by explosion.

Inner Ring: One cancer death per 100 people due to remaining radiation (5% increase): approximately 300 city blocks.

Middle Ring: One cancer death per 1,000 people due to remaining radiation (.5% increase): area similar to borough of Manhattan.

Outer Ring: One cancer death per 10,000 people due to remaining radiation (.05% increase): approximately 1000 square km

COBALT BOMB IN NEW YORK CITY (LONG TERM CONTAMINATION: CHERNOBYL COMPARISON)

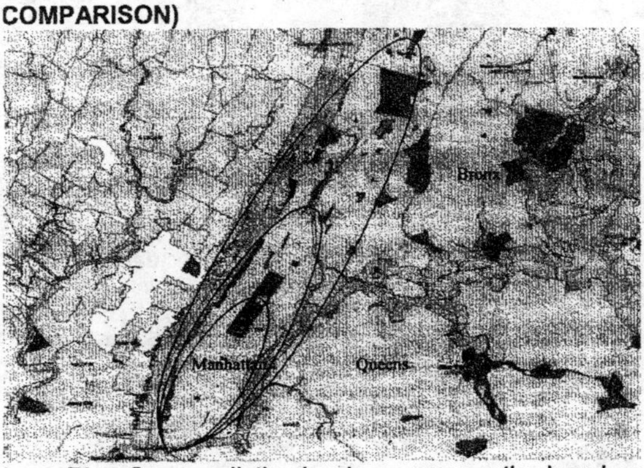

Inner Ring: Same radiation level as permanently closed zone around Chernobyl
Middle Ring: Same radiation level as permanently controlled zone around Chernobyl
Outer Ring: Same radiation level as periodically controlled zone around Chernobyl

AMERICIUM BOMB IN NEW YORK CITY (IMMEDIATE EFFECTS)

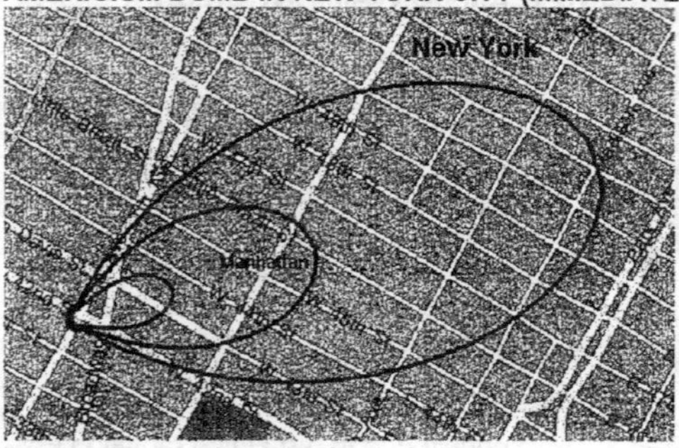

RDD: typical americium source used in oil well surveying, dispersed with one pound of TNT.
Inner Ring: Current guidelines state that all persons must receive medical supervision: 10 times area of initial bomb blast.
Middle Ring: Maximum annual dose for radiation workers exceeded.
Outer Ring: Current guidelines state that the area should be evacuated before radiation cloud passes: one kilometer long swath and 20 city blocks.

AMERICIUM BOMB IN NEW YORK CITY (CONTAMINATION)

Inner Ring: One cancer death per 100 people due to remaining radiation (5% increase).
Middle Ring: One cancer death per 1,000 people due to remaining radiation (.5% increase): approximately 10 city blocks.
Outer Ring: One cancer death per 10,000 people due to remaining radiation (.05% increase): approximately two km long / 60 city blocks.

NOTES: Above estimates might be too high by a factor of ten, or underestimated by the same factor, depending on the amount of material released, the nature of the material, the details of the distribution device, the direction and speed of the wind, other weather conditions, the size of the particles released, and the location and size of buildings near the release sight. Estimates assume a calm day (wind speed of one mile per hour), an explosion that distributes the material. Estimates do not include any direct blast injuries.

SOURCE: Testimony of Dr. Henry Kelly, President, Federation of American Scientists before the Senate Committee on Foreign Relations, March 6, 2002.

Comparing Lethal Areas of Chemical, Biological, and Nuclear Weapons

SARIN GAS

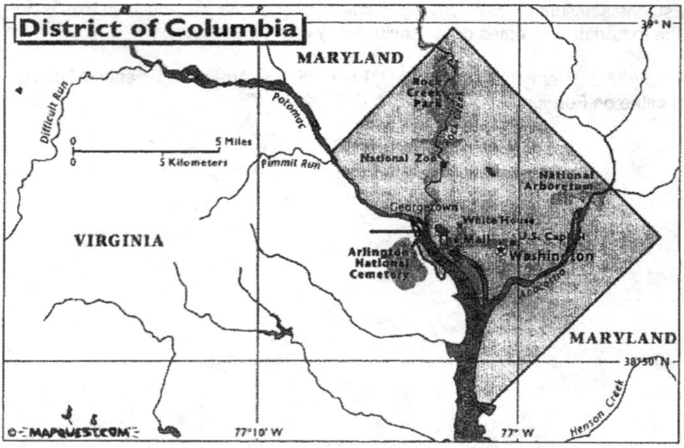

Red represents the area affected by 1,000 kg of sarin nerve gas, delivered by airplane as aerosol line source. Area affected: .8 km squared. 400 – 800 deaths.

ANTHRAX

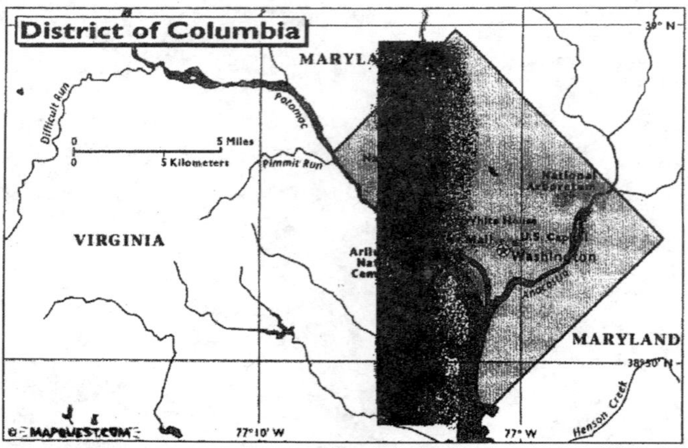

Purple represents the area affected by 100 kg of Anthrax spores, delivered by airplane as aerosol line source. Area affected: 140 km squared. 420,000 to 1,400,000 deaths.

ATOMIC BOMB

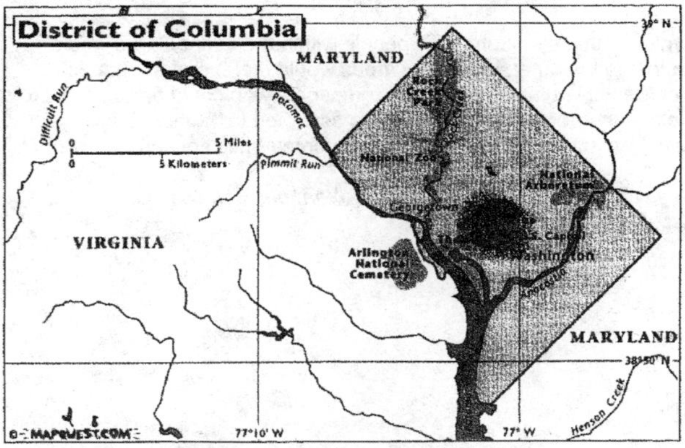

Orange represents the area affected by an atomic bomb, 12.5 kt TNT-equivalent. Area of 5 lb/in overpressure. Area affected: 7.8 km squared. 23,000 to 80,000 deaths.

HYDROGEN BOMB

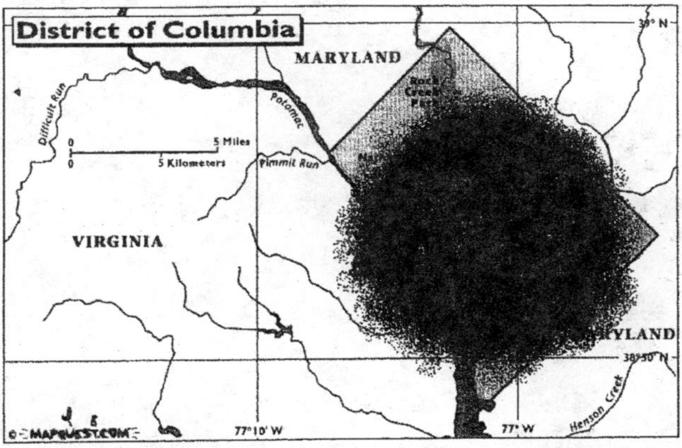

Green represents the area affected by a hydrogen bomb, 1.0 Mt TNT- equivalent. Area affected: 190 km squared. 570,000 to 1,900,000 deaths.

NOTE ON LETHALITY OF WEAPONS: Figures for sarin gas and anthrax show the lethal areas of single airplane-loads of chemical and biological weapons, assuming a highly efficient, line-source delivery of the killing agents.
The shaded areas are drawn such that although some people within the defined area would survive, about the same number in the outer, less lethal areas would die; therefore, the defined areas give approximations of the total number of unprotected people who could be expected to die in each scenario. The lethal area for the nuclear blasts is assumed to be that receiving 5 lb/in squared of overpressure – enough to level wood or unreinforced brick houses.

SOURCE: Office of Technology Assessment, *Proliferation of Weapons of Mass Destruction: Assessing the Risks*, August 1993.

Homeland Security Planning Scenarios and Summary Descriptions

Threat	Description Summary	Projected Consequences
Nuclear Denotation	Terrorists detonate a 10-kiloton nuclear device in a large city	450,000 or more evacuees; 3,000 square miles contaminated; hundreds of billions of dollars in economic impact
Biological Attack	Terrorists spray anthrax spores in a city using a concealed spray device	13,000 fatalities and injuries; extensive contamination; billions of dollars in economic impact
Biological Disease Outbreak – Pandemic Influenza	Natural outbreak of pandemic influenza that begins in China and spreads to other countries	87,000 fatalities; 300,000 hospitalizations; $70 to $160 billion impact
Biological Attack – Plague	Terrorists release pneumonic plague into three areas of a large city	2,500 fatalities; 7,000 injuries; millions of dollars in economic impact; possible evacuations
Chemical Attack – Blister Agent	Terrorists spray a combination of blister agents into a crowded football stadium	150 fatalities; 70,000 hospitalized; more than 100,000 persons evacuated; $500 million in economic impact
Chemical Attack – Toxic Industrial Chemicals	Terrorists use grenades and explosive devices at petroleum facilities	350 fatalities; 1,000 hospitalizations; 50% of facility damaged; up to 700,000 persons evacuated
Chemical Attack – Nerve Agent	Terrorists spray Sarin into the ventilation system of three commercial buildings in a city	6,000 fatalities in buildings, 350 injuries downwind; evacuations of unknown number of people; $300 million in economic impact
Chemical Attack – Chlorine Tank Explosion	Terrorists use explosives to release a large quantity of chlorine gas	17,500 fatalities; 100,000 hospitalizations; up to 70,000 persons evacuated; contamination at sight and waterways
Natural Disaster – Major Earthquake	7.2 magnitude earthquake occurs in a major metropolitan area	1,400 fatalities, 100,000 hospitalizations; 150,000 buildings destroyed; hundreds of billions of dollars in economic impact
Natural Disaster – Major Hurricane	Category 5 hurricane strikes a major city	1,000 fatalities, 5,000 hospitalizations; 1 million people evacuated; millions of dollars in economic impact
Radiological Attack – Radiological Dispersal Device	Terrorists detonate "dirty bombs" in three cities in close proximity	180 fatalities, 20,000 detectible contaminations in each city; billions of dollars in economic impact
Explosives Attack – Bombing Using Improvised Explosive Device	Terrorists detonate IEDs in a sports arena, use suicide bombers in a public transit concourse, and in a parking facility	100 fatalities, 450 hospitalizations; local economic impact; minimal evacuations
Biological Attack – Food Contamination	Terrorists contaminate food with anthrax in processing facilities	300 fatalities, 400 hospitalizations; million of dollars in economic impact
Biological Attack – Foreign Animal Disease (FAD, Foot & Mouth Disease)	Terrorists infect livestock at specific locations	No casualties; huge loss of livestock; hundreds of millions of dollars in economic impact
Cyber Attack	Terrorists conduct cyber attacks on US financial infrastructure	No casualties; millions of dollars in economic impact

Source: Keith Bea, "The National Preparedness System: Issues in the 109th Congress," CRS Report for Congress, March 10, 2005.

RESEARCH
BRIEF

Assessing the Effectiveness of TRIA

Abstract

This study simulates the expected losses from three modes of terrorist attacks to understand how TRIA would distribute the resulting losses. It finds that losses vary substantially in size and distribution by insurance line across the scenarios. It further finds that under TRIA, a large share of the losses would be uninsured and that of those losses eligible under TRIA, taxpayers would not pay for any losses from a single

After the 9/11 attacks and the substantial losses incurred, insurers questioned their ability to pay claims in future attacks and began to exclude terrorism coverage from commercial insurance policies. The fear that a lack of coverage would threaten economic stability and growth, urban development, and jobs led the federal government to adopt the Terrorism Risk Insurance Act (TRIA) of 2002, requiring insurers to make terrorism coverage available to commercial policyholders.

Because TRIA sunsets at the end of 2005, policymakers need to assess how effective TRIA is in responding to the losses from different modes and magnitudes of terrorist attacks. But because no large terrorist attacks have occurred since 9/11, no empirical data on which to base such an assessment exist.

In this study, RAND Corporation researchers respond to that need by simulating the expected losses for three different modes of terrorist attack scenarios—crashing a hijacked aircraft into a major building, releasing anthrax within a major building, and releasing anthrax outdoors in a major urban area—and then assessing the effects of the insurance system and TRIA on how those losses are ultimately distributed.

Estimating Losses from a Large Terrorist Attack

Working with the three scenarios, researchers simulated losses from each mode of terrorist attacks. Losses for the outdoor anthrax (OA) scenario dwarf those of the other two scenarios: $172 billion versus $6.7 and $7.9 billion for the aircraft impact (AI) and indoor anthrax (IA) scenarios, respectively. The distribution of those losses varies substantially. Property losses account for 13, 59, and 67 percent of the total loss in the IA, OA, and AI scenarios, respectively. Workers' compensation (WC) shows a complementary variation, accounting for 77, 25, and 23 percent of the losses in the IA, OA, and AI scenarios, respectively. Life and health lines make up 10 percent in the IA and AI scenarios and 16 percent in the OA scenario.

How TRIA Distributes Losses

TRIA applies only to commercial policies—not to life and health lines or personal lines (such as homeowners' insurance)—and covers cumulative attacks over a year. TRIA allows insurers to exclude property losses from chemical, biological, radiological, and nuclear (CBRN) attacks, but insured losses from a CBRN attack are eligible. Finally, WC losses cannot be excluded.

Some losses following a terrorist attack are "uninsured" because people and businesses do not have terrorism coverage and some losses, as noted above, are not eligible for TRIA. TRIA distributes eligible losses according to a formula. "Target insurers" who have paid claims on policies that include terrorism coverage and are in eligible lines are responsible for payouts up to an annual deductible equal to 15 percent (in 2005) of the insurer group's annual direct earned premium on TRIA-eligible lines in the previous year. They are also responsible for a copayment of 10 percent of all losses above the deductible. The federal government ("taxpayers") reimburses target insurers for the remaining 90 percent of losses above the deductible for aggregate annual insured losses in TRIA-eligible lines up to $100 billion. TRIA also requires the federal government to recoup the difference between an "insurance marketplace aggregate retention amount" ($15 billion dollars in 2005) and the sum of insurer deductibles and copayments for that year. Recoupment is collected through a surcharge of up to 3 percent per year on all U.S. "commercial policyholders."

How TRIA Distributes Losses from the Attacks

Figure 1 shows how TRIA distributes losses in the three scenarios. As shown in the black segments, *a large fraction of the loss is uninsured:* 30 percent, 14 percent, and 57 percent for AI, IA, and OA, respectively.

As the figure also shows, *taxpayers do not pay any of the losses from a single attack in any of the scenarios.* Assuming current terrorism insurance take-up rates, target insurers alone pay in the AI and IA scenarios until TRIA-covered losses reach $5 million, which occurs at $8.5 million for the AI and $6.5 million for the IA scenario. From that point on, commercial policyholders begin to contribute along with target insurers, with the two groups paying until losses reach $25 ($20) billion in the AI (IA) scenarios, at which point taxpayers would begin contributing. But since the attack losses are $6.7 ($7.9) billion in the two scenarios, at least three to four very large attacks would need to occur in a year before taxpayers would begin to contribute. In fact, assuming current take-up rates had applied then, taxpayers would not have paid in the World Trade Center attacks had TRIA been in place.

In the OA scenario, taxpayers also pay nothing, both because of low take-up rates of CBRN coverage (less than 3 percent) and because losses are spread among so many target insurers that few would meet their deductible. As the figure shows, commercial policyholders also pay nothing. Taxpayers would begin paying only after the target insurers meet their deductible *and* make their copayment, which is just above the $172 billion in losses from the attack.

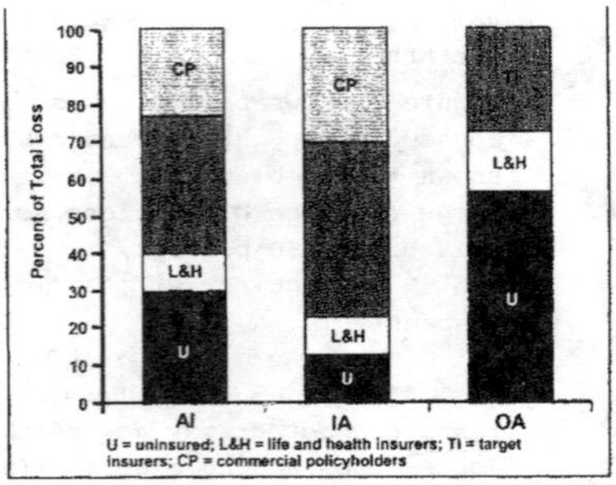

Adjusting Who Pays Under TRIA

Adjusting the provisions under TRIA can modify who pays what; however, since TRIA does not reduce losses, any modifications to reduce the burden on one group simply transfer the burden to another group (or groups). For example, one can reduce uninsured losses by extending the "make available" requirement for terrorism insurance to include CBRN coverage and by making terrorism insurance coverage mandatory. Our simulations show that doing so reduces uninsured losses but does so at the expense of target insurers, commercial policyholders, and taxpayers.

Conclusions and Implications

Based on our analysis, the role of taxpayers is expected to be minimal unless there are several large-scale attack in a single year, which means that TRIA is not primarily a taxpayer bailout of the insurance industry. Thus, modifications or alternatives to TRIA need not focus on protecting taxpayers.

Still, TRIA does influence the terrorism insurance market. Target insurers get substantial subsidies from the surcharge on commercial insurance policyholders, and TRIA makes property insurance for conventional terrorism available to policyholders, thus making terrorism insurance more available and affordable.

Also, even with TRIA, a high fraction of losses would go uninsured under the scenarios analyzed. Given this, reducing uninsured losses should be considered in modifying TRIA or in alternatives to it.

U.S. Senate Committee on the Judiciary
Subcommittee on Terrorism, Technology and Homeland Security

Director Henry R. Renteria
California Governor's Office of Emergency Services
October 26, 2005

Introduction

Good morning Chairman Kyl, ranking member Feinstein and subcommittee members. Thank you for the opportunity to testify before today on this important subject. As this brief video depicts, in California because of our many natural disasters, we, like the Gulf Coast States have learned some lessons through experience. There is no doubt that a catastrophe like Hurricane Katrina would severely tax the emergency management systems and people of any state.

We have learned these fundamental principles about disasters:

1. We cannot predict what the next disaster will bring. Each disaster has its own unique set of issues, so our emergency response system must be flexible. The answer to ten different disaster scenarios is not ten individual plans.

2. All disasters require common capabilities that must always be addressed by public safety agencies.
3. Finally, to be truly prepared for any disaster we must focus on investment in these key areas: organizational systems, training our personnel, communications and resource acquisition. These areas are critical to all disaster hazards, regardless of cause.

Systems

Because the next event--be it infectious disease outbreak, earthquake or terrorism--cannot be predicted with any true accuracy, we have learned that the best way to ensure our readiness is to develop sound and flexible systems that can be applied throughout the disaster spectrum. Ten years ago California adopted the Standardized Emergency Management System, or SEMS. As the video depicted, SEMS is a standard organizational structure used by all cities, counties, and state agencies during a disasters. SEMS also provides for standardized command and control, communications, terminology and mutual aid. Mutual aid in California is executed through a "bottom up" approach whereby a local incident commander requests additional assistance through a tiered process—surrounding local jurisdictions first, then the state, then the federal government. This is structure is commonly understood, organized, and streamlined to prevent unnecessary delay and provide access to assistance once resources are exhausted.

Also we recognized many decades ago that mutual aid between states during disaster is critical. Emergency management is fundamentally a local and state government issue. The federal government does not have the unique capabilities that states do to directly assist people during a disaster crisis. Therefore, we believe that is the best interest of the nation for the federal government to support mutual aid compacts between the state, like the Emergency Management Assistance Compact (EMAC).

In addition to the many natural and technological disasters that have tested our capability to respond and recover, California agencies have conducted numerous emergency management exercises, hundreds in just the past year.

Training

With each disaster and with each exercise we learn more and improve our capability. But the basic system remains unchanged because it has proven its effectiveness. One aspect of that system is standardization of training. My office in cooperation with law enforcement, firefighters, emergency medical providers and others have developed standardized training delivered through a training academy. In addition, each discipline, such as fire and law enforcement, has developed its own standard training that includes the common elements of SEMS.

As a result of having a standardized system, our planning at the state level has focused on assisting local governments and not preparing plans that sit of a shelf. Particular attention is concentrated on cities and counties as they primarily attend to human needs during disaster, or any emergency for that matter. We have found that a common, all-hazards planning approach is the most effective means to address the many disasters we are at risk to, whether they occur all at once or separately. In most cases the consequences of disasters will be similar; for example, an evacuation plan addressing special needs populations will apply whether there is flood or a terrorist attack.

A testament to the lessons learned and applied in California is the federal government's recent adoption of California's SEMS system, known as the National Incident Management System (NIMS). In doing do they recognize that the success of SEMS is not based upon a top-down approach to disaster management, but rather a recognition that management of disasters occurs like the disaster itself--from the bottom-up.

Communications

As we saw in 9/11 and again with Hurricane Katrina, communications between emergency responders and organizations is critical. California started developing interoperable communications many years ago to support mutual aid at all government levels; however we still have work to do to ensure the necessary communication protocols and technology are maintained and current. The Governor has initiated projects at both the state and local government level to further improve and plan for our expanding emergency communications needs. We also must ensure redundancy when traditional means of communication fail. California as a result of our experiences has implemented a number of communications means and protocols to retain contact between local, state and federal government—such as radio, satellite, phone, internet. We have also recently expanded our alert and notification system with local government, and are upgrading our statewide satellite voice and data communication service.

Resource Acquisition

Part of the communications equation in any disaster is the ability to request and acquire resources. In California, we have been fortunate to have technology in place, known as Response Information Management System (RIMS) that enables information sharing on resource needs between jurisdictions. But even this system would be severely taxed by a catastrophic disaster. The federal government should support these state developed systems that are designed to best assist first responders.

People

Finally, one of the most important lesson is that people, and not government, are the true first responders. It will take more than government systems and resources to address the most catastrophic disasters. Therefore, California has implemented initiatives such as a comprehensive individual disaster preparedness campaign and legislation which allows integration of the private sector into emergency management. The more that individuals, families and workers are trained and educated to be aware, be prepared and have skills to be and/or assist emergency responders, the quicker and more effective our response will be.

Lessons Learned

There is no doubt that hurricanes Katrina and Rita have been a wake up call to all. While we have strong emergency systems, we know that the largest scale disasters, such as these hurricanes or a catastrophic earthquake in California, will impact hundreds of thousands of people and stress our ability to preserve life and safety and recover our economy. We must also plan for the next disaster, not the last; reinforce our strengths and anticipate our vulnerabilities.

Our state and nation are rich in resources to assist in a disaster—from local government up to military assets. However, we will fail our citizens if there is not a system, organization and infrastructure in place to get this support where it is needed.

Since 9/11 we have invested significantly in emergency preparedness. But, the human toll of Katrina shows we may not have invested wisely. What can be learned from this is that the development of local emergency organizations, reinforcement of training, and investment in communications systems are what will best prepare us for the next disaster--be it hurricane, act of terrorism, or the next earthquake.

Improving Response Planning for Catastrophic Events

Statement before the Senate Judiciary Committee,
Subcommittee on Terrorism, Technology, and Homeland Security

by

Wayne C. Thomas
Vice President, Homeland Security, IEM
www.iem.com

October 26, 2005

Dirksen Senate Office Building

Chairman Kyl, Senator Feinstein, and members of the committee, it is an honor to testify today on ways we can improve preparedness for catastrophic events.

While I understand this hearing is focused on terrorism, I would like to provide a broad perspective on preparedness and technology, primarily in the context of our recent experience with Hurricane Katrina. Terrorism events and large natural disasters can produce many of the same consequences, and the lessons learned in planning for catastrophic natural disasters can be used to improve planning for terrorism events.

Our company, Innovative Emergency Management (IEM), has a unique perspective on preparing for catastrophic events. To cap twenty years of experience in disaster preparedness, we conducted the Southeast Louisiana Catastrophic Hurricane planning workshops based on a hypothetical "Hurricane Pam." Hurricane Pam was a first step toward an innovative model for integrated Federal, State, and local catastrophic planning throughout the United States. By August 24, 2005, when the fourth Hurricane Pam workshop was held, the evolving plan was still a version 1.0 with many topics still to be addressed. Katrina, which struck on August 29, 2005, required version 10.0.

Guided by this serendipitous experience, my testimony focuses on a response planning framework which produces the outcomes that our leaders and citizens demand, which is based on sound science and available technology, and which deeply engages all layers of government – Federal, State, and local. Hurricane Pam was the start of such a framework.

Define Specific, Acceptable Results

The public perception is that the response to Hurricane Katrina was inadequate. What this means is that the response did not accomplish the *results* the public expected. For example, people expected life-sustaining supplies to be delivered to those in New Orleans shelters immediately after the storm passed. During emergencies, these are the results that people care about. They are the criteria by which plans and responses are judged effective or unacceptable.

Typically, response plans are divided into emergency support functions, such as Transportation, Mass Care, and Communications. Each function separately addresses organizational roles and responsibilities, defines the overall missions to be accomplished, and identifies resources available to accomplish the missions. Piece by piece, the functional approach allocates the whole of emergency response and recovery to agencies and organizations.

What is missing is the integration of functions to focus on the accomplishment of *results*. Delivering life-sustaining supplies to those in shelters immediately after a storm (the expected result) may require coordination between several agencies and functions. If planning is focused on achieving these results, there is a greater chance the plan will prove to be effective. Plans can continue to be developed by functional areas, relying on exercises to illustrate where integration between functions is needed. However, a great deal of time and effort spent in plan development, exercise, and revision could be saved by defining the desired results for public protection first, then developing plans to accomplish those results.

This may sound like a simple concept, and in many respects it is. However, it requires policy makers, decision makers, and emergency managers to answer difficult questions about what results are desired, acceptable, and possible. In a democracy, elected leaders demand the desired, acceptable results. Emergency managers must determine if those results are possible—that is, if the results can be accomplished with available resources. Plans need to identify actions and decisions necessary for accomplishing those results.

Allow me to use a non-hurricane example to illustrate this concept. As part of a project to assess regional preparedness for a chemical weapons stockpile event at the Umatilla Army Depot, over 100 key stakeholders—emergency managers, first responders, military personnel, two Confederated Tribes of Native Americans, two states, three counties, political leaders, and Citizen's Advisory Commission members—agreed to a set of criteria defining an adequate emergency management system for a site that has been the subject of much controversy in recent years. The process took only four meetings covering a total of six days. It led decision makers to the difficult realization that they could not protect everyone. Through dialogue, decision makers defined high, but achievable results—when the emergency system accomplished these results, it would be effective. Defining and agreeing to achievable results is the foundation of a good planning process.

Community by community, this same process should be used to engage elected leadership in setting the goals for the public's protection.

<u>Base Plans on Detailed Consequence Assessments</u>

Generally, emergency response plans are not based on specific disaster scenarios. Planning annexes cover specific hazards such as industrial accidents, terrorism, and hurricanes. For example, the hurricane annex covers all hurricanes, from tropical storms to Category 5 hurricanes. However, it is normal human tendency to shy away from thinking of catastrophic events. Therefore, most plans tend to cover the last disaster that occurred in recent memory. There is not enough depth of realism in the planning basis on which specific plans are founded. There is another way.

IEM conducted the Southeast Louisiana Catastrophic Hurricane planning workshops in 2004 and 2005. These workshops were based on a hypothetical Category 3 storm called Hurricane Pam that strikes New Orleans on September 28, 2004. In these workshops, personnel from 13 parishes, 30 State agencies, and 15 Federal agencies including FEMA, the US Army Corps of Engineers, the US Coast Guard, and the National Weather Service began brainstorming and planning for a response to a catastrophic hurricane.

To create catastrophic conditions, Hurricane Pam was modeled as a strong, slow-moving Category 3 storm preceded by 20 inches of rain, spawning 14 tornadoes, and resulting in 10 to 20 feet of water within the City of New Orleans from overtopping of levees.

The consequence assessment showed that 9 refineries and 57 chemical plants shut down during the storm. The Louisiana Offshore Oil Port (LOOP), which handles 12% of US crude oil imports, shut down pre-landfall and re-opened 2-3 days after the storm.

Hurricane Pam left over 55,000 in public shelters outside Southeast Louisiana prior to landfall. The hypothetical storm displaced over 1.1 million people, affected 500,000 households and left 230,000 children out of school. Search and rescue required 20,000 boat-based missions and 1,000 helicopter-based missions for those left behind. Over 500 miles of major roads were flooded, almost 200 miles of these flooded more than 10 feet. One major bridge collapsed and almost all customers in a 13-parish area were without landline communications, TV and radio broadcast capability. Hurricane Pam generated over 12.5 million tons of debris and almost 250,000 cubic yards of hazardous household waste. Sewage treatment facilities were not working in the metropolitan area. The storm resulted in 80% of the structures affected by wind and flooding, varying from minor wind damage to total structural collapse. The total damages estimated for Hurricane Pam made it the costliest natural disaster in U.S. history. Affecting more than 12,000 square miles in Louisiana and over 1.7 million people, Hurricane Pam presented a complex web of topics and missions to be addressed in the planning for such an event.

The more than 300 Hurricane Pam planners and operators were faced with these and more specific consequences. They were facing a real storm, and they had to build action plans to respond to such an event. A catastrophic event of this magnitude would require concerted and coordinated effort from Federal, State, and local officials and agencies.

Integrate Federal, State, and Local Response Planning

At the local level, emergency responders are usually familiar with each other. They work together on a daily basis; they understand the capabilities and capacities of each responding organization and how they fit together during response. This may be a key reason why responses to local disasters are typically effective. For example, emergency personnel in county-level responses to Florida hurricanes are clearly reaping the benefits of responding to many hurricanes together.

Catastrophic events require much more than local resources. Personnel from Federal, State, and local jurisdictions must coordinate closely to ensure a successful response. What is needed is a method that facilitates collaboration and relationship-building between responders and officials at all of these levels, and among personnel from neighboring states, as catastrophes require an integrated regional and national response.

The Hurricane Pam planning process was a step toward this. The consequences of a catastrophic hurricane were presented to more than 300 Federal, State, and local personnel—both planners and operational personnel—in a scenario-based workshop environment. Personnel from different jurisdictions were divided into groups to begin tackling the operational complexities involved in addressing 15 key response and recovery results, such as unwatering the City of New Orleans, conducting search and rescue operations, establishing temporary housing, and others. The focus was on operational concerns and initiating dialogue between the many different groups involved.

Though more workshops to continue the collaboration and planning effort were needed, participants in each group were clearly focused on addressing the catastrophic consequences they had been presented with. Working together, participants developed a mission statement and concept of operations. They also identified response actions to be taken as well as available resources needed to support these actions. They were committed to producing results and there was very little finger-pointing or blame.

Catastrophes require coordinated action from Federal, State, and local agencies. For catastrophic planning to be successful, officials from all levels must be involved and committed to the process and the results. This is not always easy to achieve. There are conflicting priorities, turf issues, and resource concerns. A scenario-based planning exercise like Hurricane Pam makes the disaster real and propels officials at all levels to cut through these concerns and focus on meaningful results.

A scenario-based process can also help distinguish when a disaster is primarily handled at the local level, when it escalates into a State and local partnership, and when it requires massive Federal intervention to prevent enormous human suffering. Even in a primarily local disaster, the Stafford Act involves the support of the Federal government. If the resources of a local jurisdiction are overwhelmed, they request assistance and resources from the state. Likewise, if a disaster exceeds the capacity of the State to meet the needs of its citizens, then the state can request assistance from the Federal government. But during catastrophes, natural or terrorism-induced, the Federal government is a protector of last resort.

The National Response Plan is a good strategic document. However, an incident action plan with sufficient detail is required to handle catastrophic events impacting specific communities. Not every region is vulnerable to natural catastrophes, but some are, namely, the San Francisco region, the New Madrid Seismic Zone, Washington State, and of course, New Orleans. For these locations, a detailed and integrated catastrophic plan is the first layer of protection for saving lives.

To be effective, scenario-based plans would have to be formally adopted by all Federal, State, and local agencies who participated in their development. This is a dramatically different concept compared to current practice.

Exercise and Evaluate Against Results

Most exercises test plan compliance, checking whether responders explicitly follow the plan. However, most plans are not detailed enough and do not provide specific actions in specific timeframes. Also, most exercises do not stress the system enough, nor do they measure results in human terms. Exercise participants generally know when the exercise will be held and what will be tested. An open-book test at school, no matter the subject area, simply tests a child's ability to read. We want to test their ability to think and apply what they have learned. To truly measure plan effectiveness, it is imperative that plans be detailed and evaluation tools be designed to measure results.

Earlier, we discussed a project in which acceptable protection results were defined for a chemical weapons stockpile site in Oregon. Once these results were defined, exercises were conducted to gather data to measure performance against desired results. Where intended results were not accomplished, recommendations for improvement were provided and a "roadmap" indicating all actions needing to be completed to achieve specified results. The performance measures and action items were used by the State of

Oregon to develop work plans and budget submissions for fiscal year 2001, and the State's consistent progress toward meeting performance measures through the identification of actions, responsible agencies, and milestones for remedying areas of concern.

Citizens and their elected officials are interested in outcomes. It is hard to gauge these outcomes when evaluating either plans or exercises. One small break in the chain of response could result in hundreds of unnecessary deaths. There are many complex interactions between the hazard, the people and their unique characteristics, the shelters that protect them, the roads through which they evacuate, and many other factors. Each such interaction can end in a result that is acceptable or not acceptable.

Modeling and simulation provides a useful tool for handling this complexity. With recent advances in technology, it is possible to build a full and realistic simulation of a response and run the simulation to determine the effectiveness of a plan or an exercise for various catastrophic disasters. Improvements to the system can be modeled and their impacts on protection calculated. This provides decision makers with scientific information that can be used to make well-informed resource allocation decisions before any funds are expended. This information can also be used as the basis for resource requests.

Working together—Federal, State, and local— we can engage science and technology to build better plans and test them more rigorously. More importantly, we can produce outcomes that the public demands.

Conclusion

Our nation needs an effective method for developing plans that can be easily implemented and that deliver expected, acceptable results. The process should begin by answering a difficult question: what are the acceptable results for response? Once this question is answered, plans can be developed to and tested to see if they do so.

Planning also must include those personnel who are critical to response—responsible agencies, operational personnel, and senior leadership from Federal, State, and local levels. Commitment, coordination, and understanding necessary for successful response can only be achieved through this type of collaboration. Scenario-based planning workshops are an effective way to accomplish this.

Modeling and simulation technology that can calculate disaster consequences and measure plan effectiveness should also be incorporated into planning and evaluation. These tools offer a sophisticated method of ensuring that the plan developed accomplishes the necessary results.

The Hurricane Pam planning process was a step toward this model. A realistic, specific hurricane scenario was used to drive planning. The method emphasized collaboration between responders and focused them on specific, necessary outcomes. Clearly, the process was not perfect, but it was a step in the right direction.

For this or any type of planning process to succeed, Federal and congressional-level commitment and attention to planning is necessary. Emergency management funding has been inconsistent. State and local governments are considering layoffs in emergency management unless more funding becomes available. Many states do not even have enough funding to sustain emergency preparedness improvement initiatives they have started.

Effective catastrophic plans cannot be developed in a few months. This is a multi-year endeavor. State and local jurisdictions will need adequate and consistent funding to conduct the type of focused, intensive planning and exercising that is needed to evolve plans to the required level of maturity for an effective response to a catastrophic disaster.

We must plan together, train together, exercise together, and respond and recover together. This is the path to success.

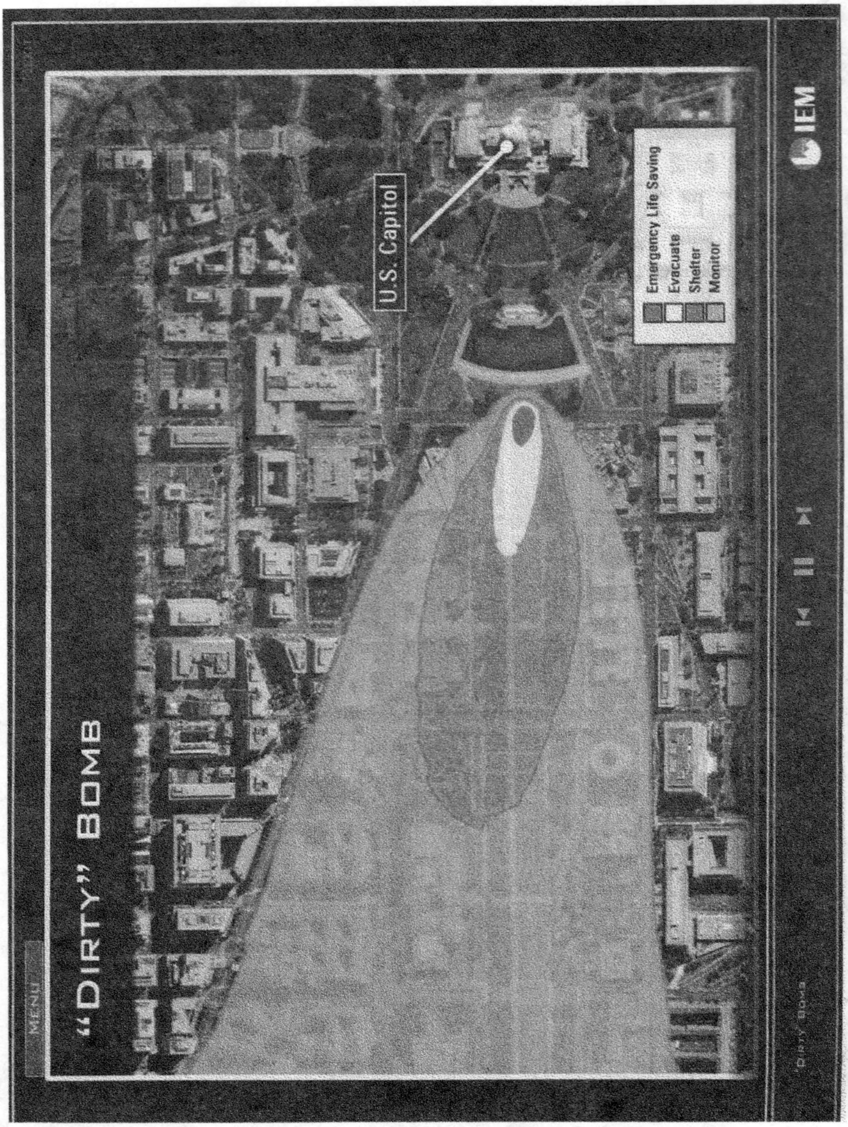
MENU
“DIRTY” BOMB
U.S. Capitol
Emergency Life Saving
Evacuate
Shelter
Monitor
IEM

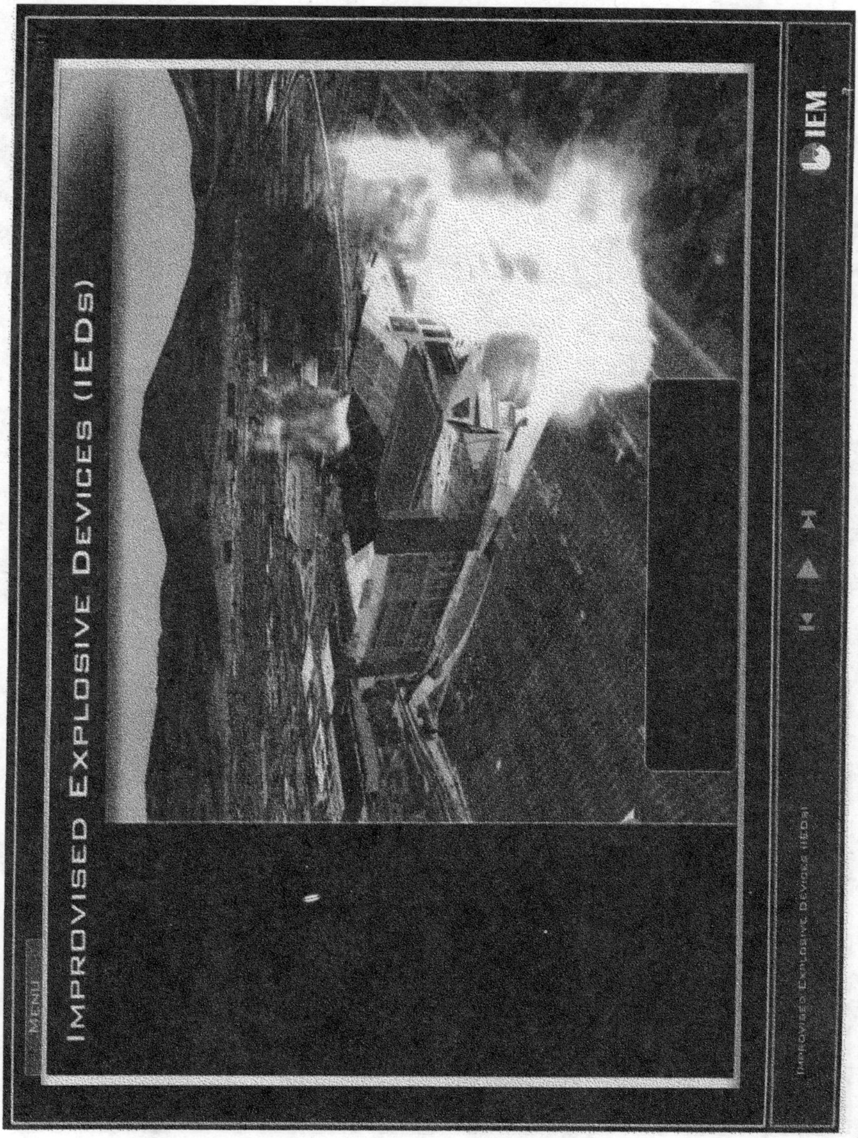
IMPROVISED EXPLOSIVE DEVICES (IEDs)
MENU
IEM
IMPROVISED EXPLOSIVE DEVICES (IEDs)

MENU
CHEMICAL RELEASE
Met Viewer
Published Runs
Latest Run / Archived Runs
Golden Gate State Park, San Francisco, CA
Plume Image
Run Summary
PAR / PAD
Edit PADs
Run Report
Print
Event ID: Normal
Run ID:
Output for: Tue Oct 11 2005, 16:45:02 CDT
Published: Mon Oct 10, 2005 16:52:08 CDT
Wind Speed (Scale) = 2 (m/s)
Risk Env.
Restricted Area
AEGL
2 - AEGL-2
3 - AEGL-3
CONTROL PANEL
D2
OPS
COM
LOG
CHEMICAL RELEASE
IEM

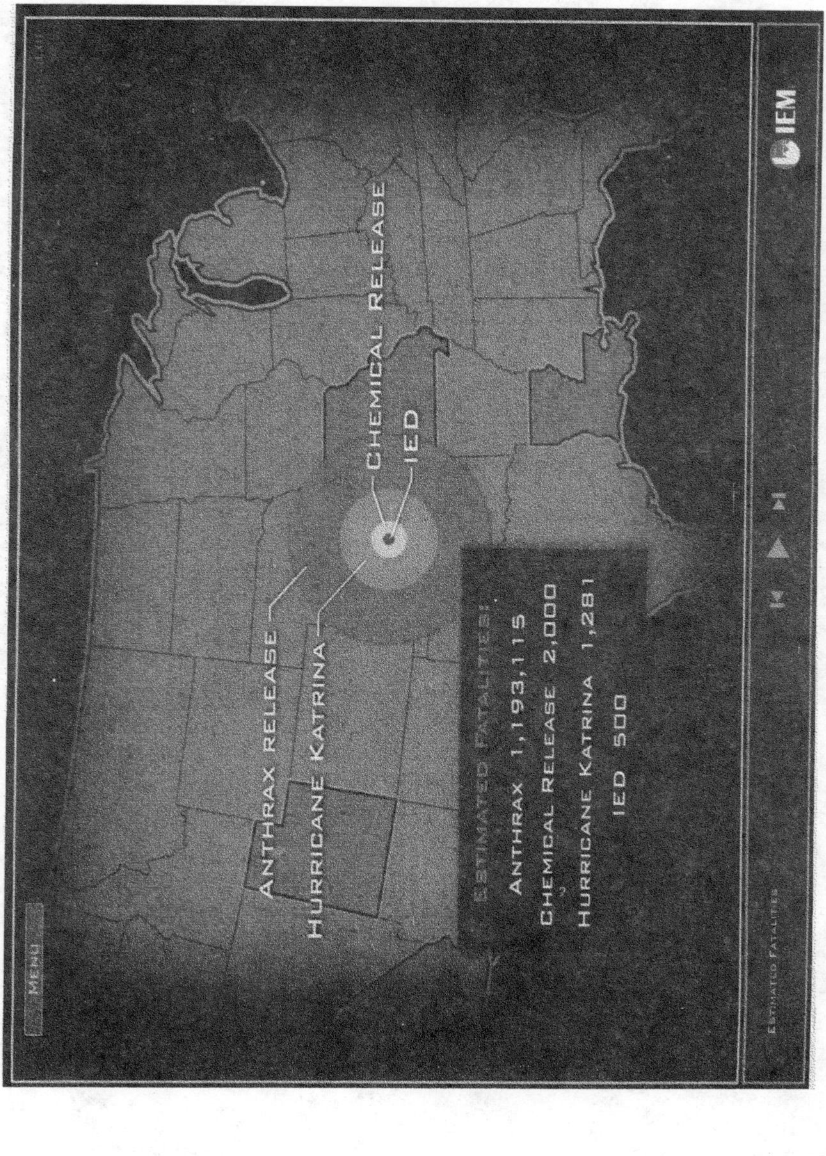
MENU
ANTHRAX RELEASE
HURRICANE KATRINA
CHEMICAL RELEASE
IED
ESTIMATED FATALITIES:
ANTHRAX 1,193,115
CHEMICAL RELEASE 2,000
HURRICANE KATRINA 1,281
IED 500
ESTIMATED FATALITIES
IEM